Legal Rights of Chemists and Engineers

Legal Rights of Chemists and Engineers

Warren D. Niederhauser, EDITOR
Rohm and Haas Company

E. Gerald Meyer, EDITOR
University of Wyoming

A symposium cosponsored by the Division of Professional Relations and the Council Committee on Professional Relations at the Centennial Meeting of the American Chemical Society, New York, N.Y., April 6, 1976.

ADVANCES IN CHEMISTRY SERIES **161**

AMERICAN CHEMICAL SOCIETY

WASHINGTON, D. C. 1977

Library of Congress CIP Data

Legal rights of chemists and engineers.
(Advances in chemistry series; 161 ISSN 0097-6156)

Includes bibliographies and index.

1. Chemists—Legal status, laws, etc.—United States—Congresses. 2. Engineers—Legal status, laws, etc.—United States—Congresses.

I. Niederhauser, Warren D., 1918- . II. Meyer, Edmond Gerald, 1919- . III. American Chemical Society. Division of Professional Relations. IV. American Chemical Society. Committee on Professional Relations. V. Series: Advances in chemistry series; 161.

QD1.A355 no. 161 [KF2940.C45] 540'.8s 77-9364
ISBN 0-8412-0357-1 [344'.73'0176154]

PRINTED IN THE UNITED STATES OF AMERICA

Advances in Chemistry Series

Robert F. Gould, *Editor*

FOREWORD

ADVANCES IN CHEMISTRY SERIES was founded in 1949 by the American Chemical Society as an outlet for symposia and collections of data in special areas of topical interest that could not be accommodated in the Society's journals. It provides a medium for symposia that would otherwise be fragmented, their papers distributed among several journals or not published at all. Papers are refereed critically according to ACS editorial standards and receive the careful attention and processing characteristic of ACS publications. Papers published in ADVANCES IN CHEMISTRY SERIES are original contributions not published elsewhere in whole or major part and include reports of research as well as reviews since symposia may embrace both types of presentation.

CONTENTS

PREFACE

This volume grew out of the recognition, some two years ago, that while hundreds of books had been written on the science of chemistry, none had appeared on the important question of legal rights and problems of chemists and engineers. The need was perceived by the staff of the American Chemical Society responsible for resolving legal problems encountered by its members. Accordingly, the Division of Professional Relations and the Council Committee on Professional Relations sponsored a symposium on this subject at the Centennial meeting of the American Chemical Society in April, 1976. The papers have since been updated for this book to include the latest legislation.

The authors represent the entire spectrum of points of view—each offers the reader new information and data on which to base his opinions and actions. Included are well-established legal precedents, legislation now before Congress which may affect the future of research in America, and suggested new legislation. Extensive bibliographies permit the reader to gain a complete understanding of those areas which he wishes to pursue further.

The audience for this publication is intended to be chemical scientists. However, research scientists and engineers in general should find the information useful. All the topics are concerned with important rights of those engaged in research and development.

The editors would like to express their sincere thanks to those who contributed to the symposium and hence this book. The contributors' efforts are an important milestone in a new attempt at awareness of individual and societal rights and of factors that will influence the future development of our technology.

Rohm and Haas Co.
Spring House, Pa. 19477
WARREN D. NIEDERHAUSER

University of Wyoming
Laramie, Wyo. 82071
E. GERALD MEYER

March 1977

1

Careers Combining Chemistry and the Law

MARCUS B. FINNEGAN

Finnegan, Henderson, Farabow, and Garrett, 1775 K St., N.W., Washington, D. C. 20006

The logic and discipline acquired in the study of chemistry and chemical engineering can be readily transmuted into the study and practice of law. Lawyers with an undergraduate degree in chemistry traditionally enter the practice of patent law. New avenues, however, have opened up, particularly with the burgeoning development of consumerism. Product liability, environmental law, and food and drug law are examples of the possibilities. The field of the transfer of technology from industrialized nations to the developing nations provides an especially important opportunity for a lawyer with a chemical background since over half of the technology transferred today is chemical technology.

We are in an interesting situation in the world today. It is a time of explosive change. Only 150 generations ago man acquired the ability to communicate from one generation to the next by recording his history, thereby preserving the accumulated knowledge of each generation and providing the means to pass it along to the next. Only 14 generations ago the printing press was invented and put into use. For the first time it was possible to disseminate knowledge generally. Only six generations ago man obtained steam power—five generations ago, transportation by railroads—four generations ago, the telephone—three generations ago, the internal combustion engine, and all it has meant to us, both good and bad—two generations ago, radio and television. Within just the past generation we have seen the evolution and general use of that amazing device, the computer, which is changing all our lives.

These achievements are evidence of the incredibly accelerated pace of change that exists in the world today. No one can afford to stand still, or he will be passed by. Over 90% of the engineers who have ever been trained in the history of the world are still practicing today. Amazingly enough, the total recorded knowledge in the world today has doubled since 1950.

The rate of growth of our knowledge and technology means both turbulence and opportunity. Because of the technology explosion, there is a larger opportunity than before for a career relating chemistry with the law. A career in chemistry, of course, represents a commitment to keeping up with the technology explosion. A career in law, on the other hand, is a commitment to ameliorate and to mitigate the displacements in our social and legal structures that are caused by the fast-paced growth of technology.

Careers which interrelate chemistry and the law are not as unusual as one might suspect. A person may obtain a degree in chemistry or chemical engineering to pursue a life-long career as a chemist or chemical engineer. Sometimes, however, he finds that a career in chemistry does not provide him with the satisfactions he anticipated. He may desire a career that involves more contact with people and less direct contact with chemistry. This sequence of changes in career objectives often leads to a desire to pursue the practice of law.

Fortunately, it is surprisingly easy to transmute the discipline of chemistry into the discipline of law. Thought processes in these two disciplines are closely analogous, and they tend to cross-support each other. To study chemistry successfully, a person must have the ability to construct a persuasive syllogism. This same ability is, of course, a primary asset to any lawyer.

Both chemistry and the law are founded on logical and symmetrical bases. Organic chemistry follows a very logical system. Certain rules must be obeyed, and if one violates these rules, he does not get the desired results. The study of chemistry teaches both deductive and inductive reasoning. A good lawyer requires both of these skills. For example, if a lawyer is writing a brief to convince an appellate court that it should reverse a decision made by a lower court, he will have to take established legal principles from earlier cases and, primarily using deductive logic, contrive them to construct a persuasive syllogism.

At the beginning of a lawsuit, a lawyer may collect many documents, mostly authored by a witness or including letters addressed to that witness. The lawyer then prepares to take this witness's testimony by oral examination (called a deposition) outside the courtroom to discover what facts the witness knows about the case. This exercise requires the lawyer to study and to analyze these documents to learn, through inductive logic, what motivates this particular witness, what prompted him to make the statements that he made in certain letters, what objectives he was seeking, and what admissions he might now make.

In a deposition by oral examination before trial, the witnesses the lawyer will be examining are normally hostile or adversary. In these situations inductive logic can be a powerful weapon which enables the lawyer to divine what kind of person the witness is, how he is likely to answer questions, what his biases are, and how the lawyer can strengthen his own case or weaken his opponent's through the examination of this witness.

Law school records show that persons trained in science and engineering typically do exceptionally well in law school. This may seem somewhat surprising, given the conventional wisdom that such major fields of study as english literature, economics, philosophy, history, and political science and government are more closely related to, and therefore better preparatory training for, the study of law than science or engineering. In actual practice, however, the training received by chemists and engineers in undergraduate school fits exceptionally well into the framework of the analytical capabilities a good law student and lawyer must possess.

One may well ask: how do you go about acquiring a law degree if you are a chemist or a chemical engineer? The major population centers of the United States—e.g., New York, Washington, Detroit, Chicago, San Francisco, Los Angeles, St. Louis—have excellent law schools. Many of them hold evening classes that a chemist or engineer can attend to receive a fully accredited law degree. In my experience some of the best lawyers with whom I have associated, attained their law degrees through evening courses. When a law student has to go to the extra trouble of holding down a job during the day, perhaps even earning his own tuition, and attending law school in the evenings, he is already demonstrating the type of ambition and determination that will make him a strong lawyer. The challenges imposed by attending evening law school tend to put a competitive edge on a lawyer that will stand him in good stead in future courtroom arenas.

Once a chemist or chemical engineer has obtained his law degree, what does he do with it? A number of different career paths are open to him. Perhaps the most traditional route would be to become a patent lawyer. Often, lawyers trained in chemistry first acquired the incentive for attending law school from exposure to patent lawyers, usually in a corporation environment. Frequently, such lawyers were actually inventors before they became lawyers.

For someone trained in the disciplines of chemistry, patent law can be a fascinating career. It is not necessary, of course, to be skilled in all phases of chemistry to be a successful chemical patent lawyer. Once one has learned the general theories of chemistry and the vocabulary, it is not difficult for inventors, scientists, or engineers to explain to him the detailed technology of their particular fields. During his career a chemical patent lawyer will probably be exposed to many different phases and nuances of the general field of chemistry. This can be a thoroughly enjoyable and rewarding aspect of such a career.

The new and burgeoning field of consumerism has opened up some new avenues for careers which interrelate chemistry and the law. Today, if a company puts a defective product on the market, people are likely to get injured, and injuries automatically create legal disputes. Those disputes are becoming increasingly complex technologically. Similarly, where environmental and ecological havoc is or might be wreaked by the careless use of technology, or for other reasons, major litigation is often the result. Such litigation can either be private or government induced.

The growth of technology has, therefore, led to branches of law which are in their early stages of development. Environmental protection law, food and drug law, consumer protection law, and others are opportunities available to the lawyer trained in chemistry or chemical engineering.

Patent law, however, is probably still the most usual area of practice for a chemically trained lawyer. This field in itself has a broad spectrum of endeavors. Basic to the practice of patent law is the preparation and prosecution before the U. S. Patent and Trademark Office of applications for chemical patents, but the patent attorney could also get involved in trade secrets, trade secret litigation, and patent infringement litigation. The patent lawyer also may deal with the important problems of licensing or transfer of technology, including the international transfer of technology to the developing or third-world nations. The chemical patent lawyer certainly has a role to play in resolving the problems which arise in the transfer of technology. Probably over 50% of the technology that is transferred today could be broadly categorized as being in the chemical field.

A law degree, however, is no guarantee that the individual will pursue these possible careers in law successfully. The question arises: what are some of the particular attributes that a person should possess to be a successful lawyer? Different attorneys may emphasize different qualities and abilities. In general, there are fairly demanding skills required in the legal profession.

Correct use of the language, especially the written language, is an important factor. Law, like chemistry, demands a specialized vocabulary. Chemists, of course, are required to use a system of chemical nomenclature which is complex but logical. It further requires a person to develop precision in using words. In chemistry, for example, there is not much difference in the spelling or often in the pronunciation of words such as butane and butene, but there is a significant difference between a butane and a butene as any chemist knows.

An attorney must likewise master the legal language, using its words and phrases precisely as well as comprehensibly. The definition of a problem, identification of alternative solutions, and explanation of the recommended course of action are major aspects in legal counseling which demand language skill. The quality of a written brief depends on the writer's ability to convey his arguments correctly and convincingly. In these instances the importance of written communication increases as the nature of the involved legal issues and problems becomes more complex, such as where law and technology intertwine.

Perhaps to a greater extent than writing skill, speaking ability is popularly associated with the attorney, and it does help a great deal if a lawyer is an accomplished speaker. In the higher reaches of the legal profession, particularly in litigation, it is necessary to be able to get up on your feet and logically and cogently to present your client's position. Regardless of whether a lawyer is involved in litigation, situations constantly arise in which he must talk to people directly and put his message across. Communication, therefore, is an essential and critical part of the practice of law. The ability to do it well is important.

Personality factors also play a role but are sometimes overlooked. Clients turn to lawyers to solve or avert problems they face, and they want to have confidence in their attorneys. An assertive, confident lawyer inspires that trust. Much of the incentive for chemists to go to the trouble and expense of acquiring a law degree arises from the desire for the increased personal contact that is offered by a career in law over many of the traditional careers in chemistry. Thus, personality is an important attribute of a good and successful lawyer.

In law, particularly litigation, many occasions arise in which the ability to analyze and to understand people becomes an important asset. In numerous situations creating a rapport with a friendly or indifferent potential witness can lead to a fuller revelation of all the facts known to the witness. If the lawyer is unable to show understanding or to relate to an individual, some facts may remain hidden.

The ability to analyze people can be especially important in dealing with hostile parties, such as key employees of your opponent. For example, in taking a deposition the good lawyer will be able to observe the witness closely and observe his reaction to the questions. The skilled lawyer can instinctively detect excessive nervousness which would characterize an area of examination disturbing to the witness. This signals the lawyer to bear down on this particular line of questioning and establish why the witness is so nervous. Pursuing the line of inquiry may enable the lawyer to get an admission on the record that will ultimately help him win his case.

Closely related to an ability to analyze and to understand people is an instinctively curious mind. Both facilitate investigation and revelation of facts—an important function of any lawyer, which like other aspects increases in importance with increasing complexity of subject matter. Chemists inherently tend to be inquisitive. This trait will help the chemist-turned-lawyer during interviewing of witnesses, examining of witnesses by oral depositions, or in open court. When cross-examining witnesses, he will follow the witness's mental processes, learn what he may be holding back, and perhaps extract testimony that will be helpful to the lawyer's case.

Another asset that chemists have when they move into the legal profession is the ability to apply themselves to a specific task—i.e., to persist until the task is properly completed, even though it may require tedious, detailed effort. A trained chemist will know, for example, when he is a lawyer representing a client in an important lawsuit, that one of the first things to do is to collect all of the possibly relevant documents. He will organize them into chronological order, study them in sequence, and painstakingly try to deduce the facts from the documents. The correspondence, reports, memoranda, and various communications transmitted between people before anyone thought about the possibility of a lawsuit, tell a story in themselves. Once the industrious and intelligent lawyer learns the facts of that story, he can effectively take testimony and cross-examine witnesses. Through a detailed and comprehensive knowledge of the documents, the lawyer will almost always be able to tell when a witness is straying from the truth because his testimony will not agree with the story told by the documents. Most chemists will have experienced similar detailed and intricate mental tasks in chemistry, involving the assembly and comparison of many facts in a logical sequence. Almost instinctively they will become good examining and cross-examining lawyers.

Another asset a chemist possesses when he enters law is a natural ability to work with and to utilize expert witnesses. In lawsuits which involve chemistry, such as patent, environmental, ecological, product liability, trade secrets, and similar litigation, expert witnesses are almost always needed. Very few judges have any extensive training in chemistry, and it is necessary to provide an expert witness to educate the judge in highly technical subject matter. The typical chemical expert has strong credentials, qualifications, experience, and credibility. If your expert is convincing in his testimony, the court will tend to accept what he says. The lawyer must prepare his expert witness for testimony, and it is extremely helpful if the lawyer himself is trained in chemistry.

A chemical background is valuable to the patent lawyer who deals with chemical technology. It gives him a foundation for talking to inventors, understanding their inventions as they are explained to him, and for asking the right questions. In many situations it may be better for the chemical patent lawyer who knows about a particular subject to act as though he knows very little in order to get the inventors or expert witness to explain the technology in his own terms. Ultimately, though, the ability to converse fluently about the technology significantly helps the scientifically trained attorney to provide the best legal advice and service.

What can the chemist who elects to pursue a career in law expect? He has many opportunities to exercise and to exploit his skills in diverse directions. He can stay enmeshed in chemistry almost to whatever extent he wishes. If he goes into chemical patent law, he can assume a position with a large corporation that has a chemical patent department. He could then become involved in a particular area of chemistry, working on a daily basis with chemists who are conducting research and development in that area. In the larger companies this could be a highly specialized area.

A chemical patent attorney could take a position with a medium- or large-sized firm and become involved in patent litigation. In this phase of the profession, the lawyer is required to match wits with an opponent in what is really an adversary contest. For those who find satisfaction in intellectual competition, this can be an exciting career avenue.

In law one gets relatively quick results as compared with chemistry. In chemistry one can work for many years on a problem before important results are realized. In law, however, most cases are resolved within two to four years after they have started. It has been said that the saving grace of law as opposed to philosophy is that in law one is forced eventually to come to a decision. Philosophers can freely speculate forever. In law one gets the opportunity to exercise philosophical skills, but the lawyer is faced with the reality that the court will eventually decide the contest. The lawyer may agree or disagree with the court, but, right or wrong, a decision is made. Even before the final judgment is reached, there are a number of interim opportunities for victory and defeat with rulings on so-called "interlocutory" matters, such as summary judgment motions and hearings on disputes that come up in the discovery or pre-trial phase of a case. A lawyer may thus have a number of minor victories and defeats as a case progresses, but when he works on something, he can usually expect a quick decision and can see the tangible results of his efforts.

Most chemists-turned-lawyers can expect to travel during their law practice. The areas of the legal profession that mix the disciplines of chemistry and the law tend to fall into the kind of practice that leads to travel, at least within the United States and often internationally. In addition this career combination will probably be financially rewarding, and more than chemistry, a legal career gives a person a chance to be his own boss. Lawyers, even in firm practice, tend to work largely on an individual basis or in small teams as cooperative units or groups. The lawyer is ultimately accountable to his clients, whom he must satisfy. Most lawyers handle their own clients, however, and in this environment the lawyer has a fair amount of freedom.

For the chemically trained lawyer, the practice of law involving technology can be intensely challenging, interesting, and absorbing. However, it is much easier to solve the problem of total immersion in legal practice than the problem of boredom sometimes resulting in other professions or disciplines. As many lawyers originally trained as chemists or chemical engineers have discovered, mixing the disciplines of chemistry and the law provides an exciting array of opportunities to achieve satisfaction in a stimulating and rewarding career.

RECEIVED September 9, 1976.

2

Proposed Revisions in the Patent Law

PAULINE NEWMAN

FMC Corporation, 2000 Market St., Philadelphia, Pa. 19103

Over the past 10 years there have been continuing and diligent efforts to change the U. S. patent laws. The major areas for which changes have been proposed are discussed, including an analysis of the changes which are receiving the most serious consideration. Emphasis is placed on the various methods for reexamination and opposition of patents, on proposals for ensuring the completeness and scientific validity of the technical content of patents, on proposals to encourage patent applicants to disclose technology ordinarily called "know-how" in addition to the technology for which the patent may be granted, and on various other proposals which are receiving substantial attention from the government and the patent community.

In recent years, there has been a sequence of bills introduced into the Congress to change the patent law. The existing law, Article 35 of the U. S. Code, was passed in 1952, coinciding with the start of a remarkable upsurge in technological growth and scientific advance.

The 1952 Patent Act worked well in this demanding environment and supported an extraordinary number of new products, new areas of business, and new businesses, large and small, all of which flourished in partnership with an effective patent system. At the same time, because of the increasing complexity of advances in technology, because of the expanding volume of scientific literature, and because of the changing methodology of research and development, certain areas of the patent law have been singled out for review and "modernization," resulting in a number of proposed bills following upon the report in 1966 of the President's Commission to Study the Patent System.

None of these proposed bills has become law partly because these bills, particularly those introduced within the past two or three years, have prompted a far-reaching debate into the role of patents in today's business and technological climate. The focus of this debate is reflected in certain specific proposed changes in the patent law, and these become apparent from a review of Senate Bill 2255 which was pending in the 94th Congress.

Many of these changes are of particular interest to chemists and chemical technology, and thus over the years the American Chemical Society has followed with interest the progress of proposed modifications in the patent law. These proposed changes have implications far beyond the purely legal/administrative patent practice. These changes affect the heart of the patent system, and the average, employed chemist has a stake in the patent system. In order to carry out research in the chemical industry for new products, improved products, and new applications and processes, almost always there is a commercial need to participate in the patent system. Without participating in the patent system, the bases for a tangible return on research and plant investment would be changed, and the incentive for innovative research and high-risk product development would be diminished.

Every country in the world has found reason to have a patent system. It is intended as an incentive system, an incentive for a major aspect of the economy: that which has to do with new products and new ideas, the commercial use of new ideas, and the investment of risk capital in new products. This incentive is more important in some fields than in others. For example, in the pharmaceutical and pesticide fields, one wonders whether there would even be private research without patents. Many people think not—or not on the present scale, but then the government, HEW, and the Department of Agriculture might fill the gap.

A patent is also an incentive for the disclosure of technological advances that might otherwise be kept secret. Patents are restricted to practical, commercial ideas; basic scientific principles are not patentable but are available to all upon their discovery and publication.

Many new businesses started with an idea and a patent. How would they have been affected by a diminished patent system? Would small inventors with good ideas merely try to sell the ideas to big business in the knowledge that without patents they couldn't compete with big business? What is the inventor's protection against appropriation of his idea while he is trying to sell it?

Yet, over the years, there have been cases where patent owners have been found to have abused their patent rights in seeking to use the patent asset for more than its limited proper legal purpose. Attempts to extend a patent monopoly to cover unpatented goods, for example, have cast a cloud over the entire system. In today's environment of free enterprise and encouragement of competition, such abuses appear to have shifted certain influential attitudes toward harsh restrictions on the role of patents in our economy. Where is the proper balance? Where is the public interest—in strong encouragement of new discoveries, or at the other extreme, a completely open marketplace at the possible expense of new discoveries? Will our national economy be stronger or weaker if we sacrifice some private research and creativity for a more open marketplace?

The answer isn't clear and is the subject of continuing and healthy debate. The position has been taken by some government spokesmen and some legislators that it's too easy today to get a patent and that corporations particularly should have extra obstacles placed in the path of participating in the patent system. Other government spokesmen, other legislators, and most of the industrial/scientific community have argued that our need for technological advance is greater than ever and that diminution of the patent incentive is not in the national interest and not in the interest of economic growth and industrial expansion.

This debate has been stimulated by the consideration in the Senate over the past few years of various proposals for changing the present patent law, culminating in the passage of Senate Bill 2255 in February 1976. This bill was not considered by the House in the 94th Congress, and it is hoped that the House Judiciary Subcommittee will hold public hearings should S.2255 or similar far-reaching patent legislation come before it.

Following are some of the proposed changes that would have an impact on the interests of chemists and the chemical industry.

Reexamination and Opposition

It is generally agreed that there should be some change in the law to facilitate public participation in the patent examination process. This is a result of the growing volume of the scientific literature and the increasing complexity of the sources to be searched. A person who knows of reasons why a patent should not have issued should be able to bring these reasons before the Patent Office, and the Patent Office should reexamine the patent and review its prior decision. These reasons are almost always published literature references—called "prior art" in the trade—that the patent examiner missed in the search.

There have been many proposals on how to accomplish reexamination. Most foreign countries have a relatively simple procedure, whereby for a few months after a patent is published some third person can file with the Patent Office, in writing, the reasons why the Patent Office should not grant the patent. The opposer and the patent applicant then argue about it in writing; if new references are cited by the opposer, as is usually the case, they argue about the references. In most countries the applicant can change the claims if appropriate to avoid the new references. Except for the possibilities of abuse in the amount of time a patent can be tied up by vigorous opposers, this isn't a bad system.

S.2255 goes far beyond this type of opposition procedure. There are two quite different proceedings in S.2255, both of which are new kinds of patent oppositions. Section 135 provides for a classical sort of opposition proceeding, available for the first year after the patent is granted but with embellishments. The opposition is not limited to written or oral arguments on a record, based on prior art or other reasons. There is available to the opposer, to the applicant, and to the Patent Office Solicitor the full sweep of federal discovery procedures—discovery of each other, of their chemists and their managers, of their files and their notebooks. There is also available, in new Section 23, the right to subpoena people and records that have no relation to either the patent applicant or the opposer and no involvement in the opposition. For example, if I at FMC wanted to oppose an application filed by Cyanamid in the synthetic fiber area, and I thought that DuPont or Eastman or Monsanto might have worked in related areas—i.e., had "prior knowledge" that might help prove that the Cyanamid invention was "obvious to one skilled in the art" or subject to other disabilities—I could seek to bring out this prior knowledge of DuPont and Eastman and Monsanto. Of course they might resist, and there would be motions to quash subpoenas, motions for secrecy orders, and many other legal actions. The Patent Office Solicitor or examiner can also do this on his own initiative. The purpose is clear and clearly stated: "a comprehensive plan for the parties to an Office proceeding to obtain evidence."

One can't argue with the philosophy behind this purpose. One can argue only with the need for so elaborate a remedy at this stage of the patent application process when the possibilities for abuse and harassment are enormous. One can tie up a patent for much of its life, which would run from the filing date and not be extended by such proceedings. This seems to me to outweigh the legitimate benefits of bringing pertinent, unpublished private information before the Patent Office to improve the patent examining process. (Published information could be submitted by simpler, standard procedures.) When you finish, if your funds hold out and assuming it's a valuable invention (if it weren't, it might not be so vigorously opposed), you may have to go through all this again in an infringement suit against the same opposer.

The chief victims of this procedure could well be small companies, or individuals, who make good inventions in fields where other companies are already established. The chief beneficiaries would seem to be established businesses who could be hurt by the competition of new ideas or improvements in their established businesses. One cannot expect that all oppositions will be filed solely with the public interest in mind. This leads to another of the objections to this sort of complex opposition proceeding: that it need not be used and there may be little incentive to use it. Thus the legitimate purpose of improving the examination of patents may be thwarted.

I am in favor of procedures that bring all pertinent information before the Patent Office. I am in favor of procedures for the citation and argument of references. Knowledge of prior use or sale should be brought out reasonably. From there on, I believe that the financial/legal burden that would be imposed by S.2255 would have an adverse effect on participation in the patent system and that this adverse effect outweighs any public benefit of not letting even one marginal patent slip by.

As a result of a lot of thinking by a lot of people, there has emerged an alternative proposal that attempts to consolidate the best of the opposition and reexamination procedures to achieve the beneficial effects and yet to reduce the costs of not only oppositions but also patent litigation. This alternative proposal has provided the focus for attempts to improve S.2255. This proposal has come to be known as "Chapter 31" because that was its place in a bill introduced in the Senate by Senator Fong. It had broad support from industry and bar associations, but it didn't carry in the Senate in its original form.

Chapter 31 provided that anyone could request the Patent Office at any time to reexamine an issued patent by citing new references. Written arguments could be submitted, the patentee could narrow his claims, and the Patent Office would reexamine the patent in the light of this new information. If, during litigation, the validity of a patent were attacked because of new references that weren't before the Patent Office, Chapter 31 required that this too go to the Patent Office for reexamination and for an advisory opinion by the examiner. This is based on the statistic that somewhat over 70% of the patents that the courts have held invalid over the past few years were held invalid on the basis of references that were not before the Patent Office, and presumably if the Patent Office had had the references, they would not have issued the patent. Reexamination under Chapter 31 would be limited to published references, and thus there would be no need for discovery or depositions or cross-examination. It would be an inexpensive procedure that would cover almost all of the reasons for invalidity that could arise in a full-blown opposition proceeding.

Chapter 31 was opposed by the Justice Department and some legislators. Observers believe that there are two major reasons. One reason is apparently that it does not allow as far-reaching an attack on a patent as could arise in a fully contested opposition, so that marginal patents or claims could slip through a Chapter 31 proceeding. The other objection is the obligatory referral to the Patent Office during litigation. It is generally believed that judges are harder on patents—especially with a vigorous adversary attacking the patent, the inventor, and the invention—than would be the Patent Office on its own reexamination. Thus, the compulsory referral of Chapter 31 was vigorously opposed.

Nevertheless, there was a partial compromise included in S.2255 in the form of a last-minute amendment that appears as Section 135A. It provides for a reexamination proceeding after the one-year opposition period has run. At any time during the remaining life of the patent, anyone can request the Patent Office to reexamine a patent based on new references. The patent owner can't change his claims as a result of reexamination, except through a reissue procedure as at present. This route can't be used if the patent is in litigation unless the judge himself decides to ask the Patent Office for an advisory opinion, but the judge doesn't have to ask for the advice, and of course he doesn't have to take the advice.

In early 1976 it appeared that the House of Representatives would take up S.2255 during that session of Congress. In anticipation of that action, Congressman Horton introduced a reexamination bill drafted by the Rochester Patent Law Association based heavily on Chapter 31 but modified in a few areas. When a similar bill was introduced in the previous session of Congress, it referred to an earlier letter to the Senate submitted by the American Chemical Society recommending stepwise legislation in specific areas of the existing patent law and suggesting that opposition proceedings would be a good place to start. In June 1976 Congressman Wiggins introduced a bill which embodied Chapter 31 in its original form as an amendment to the existing patent law. This approach has received general support from the patent community as a solid and important step in meeting the needs of a patent system intended to encourage technological growth.

Joint Inventorship

Of major concern to the scientific community are those aspects of S.2255 that relate to joint inventorship. Legislation is needed in this area to clarify ambiguous and conflicting decisions, law, and practice. S.2255 took a giant step backward. With the growth of technology and the increased complexity of inventions, often more than one person makes an inventive contribution to a patentable advance. Not every invention is created full-blown in the mind of one person but is creatively developed, sometimes by teams of researchers, sometimes by successive contributors. S.2255 expressly does not permit this recognition of how inventions are made. S.2255 requires that all joint inventors must have contributed to every claim in the patent. Proponents of this requirement have stated that their purpose is to reduce, as a matter of national policy, the issuance of patents based on "corporate inventions." I do, indeed, think it will have this effect. It presupposes that inventions made by two people are not in the national interest while inventions made by one person are. I know no basis for the conclusion that complex invention and corporate technological leadership in this country should be isolated from our only invention incentive system. The existing law on joint inventorship needs clarification. Other bills before the Senate offered advances in this area, but these were not embodied in S.2255.

Assignee Filing

S.2255 provides that a patent application may be filed and issued in the name of the patent owner, provided the inventors are correctly identified and provided that joint inventions meet the requirements discussed above. This is different from the present law, which provides that a patent is filed and issued in the name of the inventors with the patent owner also listed on the patent. I know of no corporate group that vigorously urged this change, and it provides very little legal advantage. If it was inserted as a sop to the corporate applicant, it is a minor concession. Ownership rights as to patent applications are based on the legal relationship between the inventor and the assignee and not on the technicality of in whose name the patent application is filed.

Disclosure Requirements

There seems to be, in some circles, the suspicion that the draftsmen of patent applications try to conceal the substance of an invention rather than to emphasize it. Whether this was ever the case, I'm not sure, but today the penalties for providing anything other than a fully enabling disclosure are so great that a patent applicant is certainly ill-advised to play that game. One hears chemists complain about the difficulties of trying to repeat experiments in the *Journal of the American Chemical Society*, or worse, in *Chemical Abstracts*, because of the lack of detail. Most patent applicants in the chemical field find today that it is advisable legally to include more detail than would be included in a journal article simply because the risk of even an appearance of withholding pertinent data carries such high penalties. S.2255 provides for technical disclosure and review and discussion of the literature well beyond that which a chemist might feel called upon to include in a technical article on which his professional reputation might ride.

There are new provisions in S.2255 relating to the obligation to make "reasonable inquiry" into all related information "in the possession or control" of the inventor, the applicant, the assignee, and the patent agent or lawyer. Some people interpret this as meaning that a lawyer with different clients might be compelled to tell the Patent Office the trade secrets of one client if they might have any relationship to the patent application of another client. This is perhaps not the intention, and I think it could be clarified. It is generally agreed that this clause, although not unreasonable on its face, provides further gimmicky technicalities to be resolved in future litigation.

Importation of Products Made Abroad

Section 271 of S.2255 would provide some small measure of protection to manufacturers when their process which is patented in this country is practiced outside of the country, presumably by cheaper labor, with the goods then imported to the detriment of the U.S. manufacturer and to labor paid on U.S. standards. However, this provision takes effect only if the importer is the exclusive or primary distributor. If there are several non-exclusive distributors—none being "primary"—this safeguard would not apply. This entire provision thus would be easy to avoid. There are important issues here involving labor policy and international trade as well as patent law and fairness. Any change in the law should consider all the issues.

Another new provision in Section 271 provides that a patented invention, if substantially completed within the United States and then finally completed elsewhere, cannot avoid infringement of the U.S. patent. This is a useful clause because a recent U.S. court decision had held that all aspects of a patented invention must be practiced within the United States in order to infringe the U.S. patent. It now remains for the courts to decide what is meant by "substantially completed."

Patentability Brief

Another provision of interest to chemists is the compulsory filing of a "patentability brief," wherein the inventor discusses pertinent literature references and other background information and explains why his invention is patentable in the light of this background. This requirement for a patentability brief is not in itself disadvantageous. It does, however, present risks to the patent applicant when considered in the context of other provisions of S.2255, particularly that which requires that the inventor, the assignee, and the attorney investigate all sources of information within their possession or control—this surely means within all laboratories, even overseas laboratories of a multinational corporation, and all other chemists working in these laboratories—to include in the patentability brief all pertinent information which the company may have. Any chemist in a big company who thought he was working in isolation will now find himself in contact with colleagues throughout the company and its subsidiaries.

Deferred Examination

Deferred examination was initially used in countries with five or more years of backlog of untouched patent applications and where the backlog was getting worse each day. The U.S. Patent Office had a large backlog itself a few years ago, but thanks to efficient commissioners and various procedures for accelerating prosecution, at present in the United States the majority of patent applications are processed within 18 months of filing. Nevertheless, S.2255 provides for deferred examination, even though few patent users are now urging this step.

There isn't time to go into the arguments for and against deferred examination or to discuss the other new provisions contained in S.2255. Many of us who believe that a good patent system, designed to encourage technological progress, is important to our country, are concerned about the major and minor changes being proposed in the patent law with inadequate study and inadequate public participation.

Cost and Benefits

How much should it cost to get a patent? Who should pay this cost: the inventor? the government? the public? An Inflationary Impact Statement was prepared by the government in connection with one of the bills that led to S.2255. It was estimated that the cost to the government under present law averages $1500 per patent application (this average was based on the total of mechanical, electrical, and chemical patents). The statement estimated that an additional $1233 per application would be added, making the cost to the government an average of $2733 per application. The government also estimated that the increased cost to the patent applicant would be 80% over present cost. Many industries have estimated the increase to be several times that amount, with the highest estimated increase coming from the chemical industries because chemical processes and the ordinary practice of chemical experimentation would present the greatest burdens in complying with the proposed new law.

The government, using these figures, estimated that the rate of patent filing would drop by one-third. They did not respond to the question of whether this is a desirable result or the desired result. Nothing has been heard from the sponsors of the administration-supported patent bills as to whether this result is in the national interest. Perhaps we really are better off with one-third fewer technical disclosures, but which third? Many of our major industries, particularly high-technology industries, are deeply involved with the patent system. We don't know how the next generation of possible new industries would approach the high investment and high risk climate of today. The consequences are not well understood and concern us all.

Summary

S.2255 represents, I believe, a calculated move toward a diminished patent incentive system. If this philosophy prevails—that patents should indeed be a diminished factor in our competitive economy—we may never know where our technology might have gone in an environment more supportive of creativity and new ideas. It is my personal view that the risks of damage to our technological future are sufficiently real that no change of the magnitude of S.2255 should be made unless we have a better idea of the consequences. I am not in favor of S.2255 because I believe that the disadvantages outweigh the advantages. I support stepwise amendment to the present law, to modernize it where appropriate, to codify judge-made changes in the law as appropriate, to clarify ambiguities that have developed since the 1952 Patent Act, and, as the foremost consideration, to provide an increased incentive to our national economic strength and technological preeminence. I hope that scientists and chemists, as users of the patent system, will speak out on whatever their views may be as new patent legislation is proposed.

RECEIVED September 17, 1976.

Discussion

Q. What can the average person do in this context?

A. I believe that participation in the political process is in order—to express whatever views a person might have. I think that is about all the average person might do, but it is something that we are inclined not to do, and I think this legislation is important enough to act on.

Q. What do you think will be the cost to the average small inventor under this new bill?

A. You have to look at the cost of this bill in phases. The filing phase will probably increase his legal fees substantially in terms of the increased effort in filing and prosecution, but the real cost and, I think, the real hazard is the possibility of putting the small inventor in a situation which he can't handle financially in the opposition aspects of the bill; these come at a time fairly early in the life of a patent when the inventor may not know the true worth of the invention. The diversion of the inventor's resources to what has been compared with a full-scale district court proceeding, with the kinds of discovery and testimony that one sees in patent litigation, would put the patent opposition system in the context of litigation rather than in an administrative proceeding. The figure of $10,000—20,000 is being used in evaluating the cost of fighting an opposition through its full potential burden under S.2255.

Q. Is there any chance of amending the bill to separate the corporate inventor from the private inventor?

A. There has been an attempt in the bill to recognize the disabilities, the extra burdens, on the small inventor in that there is a provision which says that for an individual inventor, or those who meet the definition of a small business as defined in our laws, there is an upper limit on the filing and issuance fees of the patent. That upper limit is $100. That same clause says that there will be a minimum lower limit of $200 for the corporate applicant. Now that difference does not solve the question that you raised, but it is as far as the sponsors of the bill are apparently willing to go. I think that when associations of big business tell the government that they are worried about how this is going to help or hurt the small inventor, they don't receive much attention. It isn't remembered that most big business started small.

3

Special Compensation for Salaried Chemists and Rewards for Inventors

WILLARD MARCY

Research Corp., 405 Lexington Ave., New York, N. Y. 10017

Special compensation for the employed inventor has been mandated by statute for many years in a number of countries foreign to the United States. In this country such compensation, if any, is customarily provided by individual employers at their discretion. While a bill to provide mandatory compensation has been introduced with modifications by Rep. Moss of California at each session of Congress since 1970, no action has yet been taken. During this period the ACS Committee on Patent Matters and Related Legislation, jointly with the Economic Status Committee, has been studying the various provisions of the bill in order to adopt an official ACS position on it. The relationships between this bill and existing legislation in other countries are discussed. Proposals and procedures other than legislative are outlined for on-going discussion.

Paradoxically, the world's first society to recognize through law the right of employed inventors to receive a reward for their work was Hitler's Germany. The country's desperate situation in the early 1940s called for the immediate development of new products and processes, and it was proposed that inventors and innovators should be stimulated by monetary rewards proportionate to the value of their contribution. Thus was born the German "Law Relating to Inventions of Employees" which went into effect on July 21, 1942.

Today, we also have an urgent need for innovation, albeit for vastly different purposes. Should we stimulate inventors through a legally determined method of compensation, and if this is impractical or undesirable, what other methods can be used to increase creativity?

Since World War II public interest in innovation has been steadily increasing. Conservatively, innovation is a growth industry. While the words were scarcely known and seldom used 25 years ago, talk of "innovation" and "technology transfer" can now be heard at almost every national and international conference, and the words can be read every day in lay and technical journals around the world.

"Innovation" has become a catchword for describing a highly complex process. Strip away the superficial appearances of common usage, logic, and good reasoning, and you find strong biases, illogical opinions, emotions, and legalisms. For this paper, however, let us be content with a simple definition. Innovation is the procedure by which new products and processes are conceived, developed, and introduced into public use. Innovation is a simple concept in theory; in practice, it is extremely complex.

Why is innovation important, and why are so many people interested in it? One reason is that accelerating progress over several decades now leads people to ask: "What's new? What's better? What will make us healthier and happier? How can we protect our environment and expand our sources of energy? How do we maximize the good innovations and minimize or eliminate the bad? Are public funds, increasingly devoted to innovation since World War II, being well spent?"

Returning to our original question concerning compensation for employed inventors, some assumptions are necessary before we can attempt an answer. We assume that change is necessary and desirable, and the innovative process cannot be stopped. We assume at least part of the innovative process must be carried through before we can judge whether any given development is good or bad. To ensure that we do not miss the good innovations, we assume that we must encourage and promote the use of the innovative process generally. Finally we assume that motivating individuals to invent is one way to produce more innovations.

Based on these assumptions, and taking human nature into account, an obvious conclusion is that rewarding inventors will motivate them to invent. Man-on-the-street inventors, if successful, gain their rewards directly from public use of their inventions. Employed inventors, however, must depend on the goodwill, imagination, and largesse of their employers—industry, government, or academia. Although many large corporate employers do have voluntary employee–inventor compensation programs, goodwill and largesse may not be forthcoming from many, if not most, intermediate and smaller companies and government employers without some direct or indirect persuasion.

While Germany operated under the "Law Relating to Inventions of Employees" beginning in 1942 (rewritten in 1957 and amended in 1961 and 1968), other nations either did not know of the practice or did not recognize its effects until well into the 1950s. Since then at least 15 countries have developed and promulgated similar laws based on the German precedent. All stem from the assumptions that monetary stimulation will motivate inventors and innovators to the public benefit and that the adequacy of such stimulation depends on the force of law.

No similar legislation was introduced in the United States until Rep. Moss of California filed a bill in the U. S. House of Representatives in 1970. His original bill followed the format of the German law with some modifications to make it more applicable to conditions in the United States. The proposed legislation, filed primarily at the instigation and with the help of the Coordinating Committee of the California sections, a coalition of American Chemical Society sections, and certain other professional societies, had no congressional action taken on it and expired with that session. New bills with modifications were filed in subsequent years, the latest dated March 25, 1975. No action has been taken on any of these, and no corresponding legislation has been filed in the U.S. Senate. The 1975 version was still pending when the 94th Congress adjourned.

A comparison of the latest version of the Moss Bill with the current German law shows certain similarities. Both relate to rights in patentable inventions and proposals for technical improvements, and each applies to all types of employed inventors—industrial, academic, civil servants, and armed services. Both distinguish between "service" inventions, to which employers have rights, and "free" inventions, which belong to the employee. Each contains general language regarding proper and adequate compensation for employed inventors. Both include provisions relating to domestic and foreign patenting and providing for internal counseling and for outside mediation (arbitration) and legal procedures in cases of unresolved controversy.

Definitions of technical improvements differ. As compared with Moss Bill, the German law places greater restrictions on the rights of inventors employed by government and academe. The German law provides that the employer can obtain nonexclusive as well as exclusive rights; the Moss Bill is limited to exclusive rights. Differences exist as to the extent of rights in foreign patents acquired by the employer. Differences also exist in the internal counseling procedures and in patenting procedures related to differing patent laws in the two countries.

In order to execute the provisions of the German law, directives have been issued which, while stated to be guidelines, have, in effect, the force of the law itself. These directives spell out procedures for determining the compensation to which inventor-employees are entitled. The factors taken into consideration are the value of the invention, the employee's duties and position, and the contribution of the employer to the making of the invention. Complex formulas and tables are provided to aid in assessing these factors. No similar directives or guidelines have as yet been proposed in connection with the Moss Bill.

When the first Moss Bill was filed, its provisions were studied by the American Chemical Society Committee on Economic Status and the Committee on Patent Matters and Related Legislation. One concern was whether an official ACS position on the bill should be presented to the governing congressional committee. The ACS bodies also felt that consideration should be given to the active involvement of ACS itself in developing a program for encouraging compensation to employed inventors. It soon became apparent to the members of the two ACS Committees that the issue was more complex than appeared, and a joint subcommittee was appointed to study the entire question in depth. This subcommittee, under my chairmanship, has been active since 1972, and it made a report to its parent committees in the fall of 1974 with recommendations for future action.

An initial effort by a prior subcommittee of the Committee on Economic Status, involving a survey of a representative number of industrial companies, had developed details on the then-current methods for compensating employee-inventors. The joint subcommittee, however, felt that recommendation of a formal position to be taken by ACS would first require answers to some basic questions, for example:

- In terms of the public interest, what innovations are needed?
- How do we get inventors and innovators to contribute to the satisfaction of these needs?
- What parties would be likely to have interests in such innovations and of what nature?
- What are some alternative means which can be used to provide these stimuli and rewards?

To obtain some understanding of these basic issues, the joint subcommittee decided that broader surveys were needed, particularly within the ACS membership. Such surveys, however, would be prohibitively expensive and time-consuming even with efficient sampling techniques. Instead, the two parent committees were persuaded to agree to sponsor jointly a Public Hearing at the fall 1973 ACS National Meeting in Chicago. A printed, edited transcript of this meeting is now available from ACS Headquarters.

The ACS hearing produced some guidance and even some answers to the basic questions but provided little insight into how the existing laws have worked in foreign countries, particularly West Germany. To overcome this, the Committee on Patent Matters and Related Legislation sponsored a meeting in San Francisco in May 1975. The meeting presented and discussed input from foreign attorneys who have practiced for long periods under existing compensation laws. An edited transcript of this meeting is included in the same ACS document as the Public Hearing transcript.

After studying all of the information developed to date, some conclusions can be drawn which will guide future ACS activity. While the conclusions which I present are my personal views, I believe the joint subcommittee members share them, perhaps with minor differences of opinion.

(1) There has been sufficient interest by the ACS membership in this issue and enough expressed dissatisfaction with many existing corporate compensation plans to warrant continued work to develop and to execute an action program by the Society.

(2) There is no real opposition by corporate management to providing extra compensation to employed inventors.

(3) The principal problems to be resolved in providing such compensation are to determine what constitutes "fair treatment" of employed inventors, to determine who should share in any compensation, and how to determine the value of an invention. These are complex problems that involve not only many technical and financial decisions but emotional and psychological considerations.

(4) Aside from legislation there are alternate methods of providing reasonable additional compensation, and these should be explored since they might be preferable.

(5) Experience under the German law and similar laws in other countries indicates that such laws are workable, but the cost of administering them is substantial. Neither employed inventors nor employers are completely satisfied with them.

(6) There is no unequivocal evidence that such laws actually stimulate either the evolution of new and useful inventions nor their introduction into the marketplace. On the contrary, there is some evidence that they have had the reverse effect by stimulating research workers to maintain silence.

(7) Except for a few special instances, there is little evidence that providing an employed inventor an opportunity to exploit his inventions himself, if his employer chooses not to do so, has been a major benefit either to the inventor or society. In almost every case such inventions never get off the ground for economic reasons.

I suggest a course of action which does not require immediate legislation and which can be undertaken in a professional manner by one or more professional societies. This proposal is made on my responsibility alone, and does not necessarily reflect the opinions of other subcommittee members, nor does it have the endorsement of either of the parent committees or ACS itself.

Laws and a legal structure are essential in criminal matters or in blatantly unfair and exploitive civil situations. In matters of honest differences of opinion, I find it abhorrent to rely solely on legislation and the strong arm of the law to force a resolution. While it is necessary to define in law certain principles or limits, the legal demarcation of each step in an action to resolve differences of opinion is undesirable, unnecessary, and counterproductive.

Furthermore, the amicable resolution of differences requires a convergence of views arising from a better understanding between individuals and organizations. To attempt such resolution using adversary proceedings, as are inevitable under strict laws, can only prolong and inhibit final agreement and cause intractable animosities. This is illustrated by experience in those countries that have compensation laws.

The following sequence of actions by ACS would be a more fruitful way for the Society to proceed rather than to rely solely on responding to proposed legislation. The ACS should:

(1) Include in its "Guidelines for Employers" strong statements that tangible awards should be provided for specific contributions by employed inventors; further, inventive discoveries of no interest to employers should be released to employees. The ACS should encourage by direct communication and collaborative action the inclusion of similar provisions in employer guidelines published by other professional societies.

(2) Develop typical plans for compensating inventors for use by employers and encourage all companies, government agencies, and educational institutions to administer such plans.

(3) Establish an office at ACS Headquarters to help individual employers set up equitable compensation plans for employed inventors. Such a service could be either self-sustaining through a fee system or provided at ACS expense. Joint efforts along these lines might be taken with other professional societies.

(4) Establish within ACS a counselling, mediation, and conciliation service for the benefit of members and employers in resolving issues relating to compensation for employed inventors. This service could be gratis or for a fee and could be a joint undertaking with other professional societies.

The program outlined here avoids the involvement of new government bureaucracies and government intervention in essentially private matters between two (or more) parties. The bargaining strengths between the parties are put on a more equitable basis, and moral suasion and economic pressures can be more effectively brought to bear on both employers and employees. Expensive legal proceedings are avoided.

It is possible that additional pressures may be needed to arrive at a satisfactory system, and some form of legislative backup may be desirable. Any law enacted, however, should be much simpler than either the German law or the Moss Bill. Such legislation should relate only to providing legal means to resolve otherwise unresolvable situations. It should not spell out in exhaustive detail the rights of the parties and the administrative procedures needed to implement these rights.

Although much progress has been made in the past five years in recognizing and understanding the problems inherent in compensating employed inventors, effective action has not yet been taken. The joint subcommittee expects to continue its work and to arrive at a recommendation for an official position on the Moss Bill. However, such a recommendation will not be made until it is apparent that the bill will be scheduled for a congressional committee hearing. Meanwhile, a program such as that outlined above appears timely. It is hoped that the ACS will take the initiative.

Bibliography

Brennan, J. W., "The Developing Law of German Employee Inventions," P.T.C. *J. Res. Ed. (IDEA)* (Spring 1962) **6** (1).

Cartright, H., translator, German Law Relating to Inventions of Employees and Directives Issued thereunder, 2nd ed., Uexküll, D., Königgrätzstrasse 8, Hamburg, Germany, 1971 (English translation).

Conner, M., "The Moss Bill," *ChemTech*, August, 1972.

Harter, F. C., "Statutorily Decreed Awards for Employed Inventors: Will They Spur Advancement of the Useful Arts?" P.T.C. *J. Res. Ed. (IDEA)* (Winter 1971–72) **14** (4).

Lassagne, T. H., "Analysis and Critique of Moss Bill, HR.15512, 91st Congress," Report to Committee 106, Inventors, of the Section of Patent, Trademark & Copyright Law of the American Bar Association (1970).

Moss, J. E., "HR.5605, A Bill to Create a Comprehensive Federal System for Determining the Ownership of and Amount of Compensation to be Paid for Inventions Made by Employed Persons," Introduced into U.S. House of Representatives, 94th Congress, First Session, March 26, 1975.

Neumeyer, F., "The Law of Employed Inventors in Europe," Study for Subcommittee on Patents, Trademarks, and Copyrights of the Committee on the Judiciary," U. S. Senate, Study No. 30 (1962).

Quigley, S. T., "Perspectives on Inventor Compensation," Symposium on Patent Awards for the Employed Inventor, 70th National Meeting, American Institute of Chemical Engineers, August 30, 1971.

Röpke, O., "Der Arbeitnehmer als Erfinder seine Rechte und Pflichten—Ein Practischer Ratgeber" (1966).

Sutton, J. P., "Compensation for Employed Inventors," *Chem. Technol.* (Feb. 1975) p. 86.

RECEIVED August 10, 1976.

Discussion

Q: What about all the people necessary to make a successful invention. How are they going to share in special compensation?

A: What you are asking points up that this is not a simple inventor compensation problem. It is a much more complex situation. If an employer develops a plan for compensation, he should take possible complex situations into account. This is the gist of what I am saying.

Q: What kind of examples of extraordinary compensation are you talking about?

A: I hesitate to get into that because there are three ways that are set forth in the German law to compensate the inventor. One is by using an analogy to licensing. If there were a license issued for an invention at a certain royalty rate, then a percentage of that royalty would be paid back to the inventor as his share. Another way is to try to determine the value of the invention in terms of profits back to the company and take a percentage of that. The third way is to simply come to some mutual understanding between the employer and inventor(s) on an arbitrary basis.

Q: Have you made any specific recommendations to the Committee for adoption?

A: Yes, the first point in my action plan was that two guidelines have been recommended by the Committee on Patent Matters and Related Legislation to the Committee on Professional Relations for inclusion in the ACS Guidelines for Employers. One has already been adopted by the latter committee, and the other is being considered for adoption at this meeting. The one that has been adopted reads this way: "tangible awards should be provided for specific contributions by employed inventors." It should appear in the next edition of the Guidelines. The second one, that has not yet been adopted, is that inventions or discoveries of no use to the employers should be released to the employee.

4

Confidentiality, Secrecy Agreements, and Trade Secrets

S. BRANCH WALKER

American Cyanamid Co., Stamford, Conn. 06904

Industrial property is best protected at times by confidentiality or by a secrecy agreement as a trade secret. The terms have somewhat different meanings but overlap in part. In general, high ethical standards in the profession are the best guide for proper conduct. The chemist, the employer, and the public all need to be considered and protected. The relationships between a former employer, a current employer, and a chemist need to be considered carefully to separate proprietary data of the first employer, not to be disclosed to the new employer, from what is the professional skill and knowledge of the employee, which are properly available to the new employer or prospective employer. Written agreements help to interpret the rights and duties of all parties. A few typical examples are cited to show court rulings.

In common with most other papers on controversial subjects, any opinions expressed herein are not necessarily those of my employer or of any organization or group to which I belong. Some of the opinions are quotations from reputable sources with which I may not even agree. The opinions may also be inconsistent.

Legal differences of opinion, particularly on controversial subjects, are not uncommon. In a very recent decision of the Supreme Judicial Court of Massachusetts, for instance, the seven justices filed six separate opinions on a mandatory death penalty, holding the statute unconstitutional (*1*). Perhaps the number of dissenting or specially concurring opinions shows that only controversial cases go to the highest courts, and we need the clarity of thought that goes into their well written opinions. We might too consider that before there is a case before the courts at least two parties must have a serious difference of opinion, a belief that the court will hold in their favor, and a substantial bankroll to finance the litigation.

In discussions of trade secrets and associated agreements on confidentiality and secrecy, we can only try to forecast the future based on the record of the past, largely as expressed in court opinions, on the relationships that have grown up based upon what parties think is right, and on what the parties think a court would hold if a question were put to it.

Many people wonder why laywers are often so prolix. Usually it is an attempt to be clear. It is quite common for a lawyer to restate a question together with the answer. It is not at all uncommon for a client to present a long involved set of facts on which he wants a "yes" or "no" answer only to get a good many pages of restatement of the facts before a conclusion is reached. This is necessary to ensure that the client and the lawyer are discussing exactly the same question. Closely related questions may have different answers, and the client may bias the question so as to receive the response he wants. In such biasing, the question can become sufficiently different that it does not fit the circumstances found later to exist.

One classic case in the legal profession is the story of a driver who described a collision at a particular intersection and asked the lawyer's opinion on whether or not he was negligent and should be held accountable for the damages. The lawyer said no, but the driver forgot to mention that there was a stop sign barring his entry into the intersection. This change in facts destroyed the pertinence of the opinion.

Hence it is first necessary to consider that factual situations vary. Then one must consider that a lawyer can state how a particular court in a particular set of circumstances set forth the law and from this can give an opinion estimating what the law is today. A top-notch lawyer can make a reliable estimate of what the law is going to be when a future dispute is adjudicated in a special tribunal. Also, a great part of a lawsuit is establishing facts, and this is particularly pertinent to trade secret situations.

The 10 Commandments clearly and unequivocally state "thou shall not kill." Most of the statutes on the point are remarkably longer. In most instances the real question for the court is not whether someone was killed but what admissible facts can be presented to establish that a particular individual is responsible. A great many trade secret problems involve differences of opinion as to what the facts are. If we are attempting to adjudicate the relative positions of parties in situations concerned with trade secrets and secrecy agreements, we are more apt to run into questions of differences of opinion over what is a trade secret and who compromised it than in determining the elements of the basic law. The real problem is to apply the law to the facts.

There are at least four sets of "facts": (a) the facts as the plaintiff sees them, (b) the facts as the defendant sees them, (c) the facts as the court and jury sees them, and (d) the facts as they really exist. In attempting to determine the facts there are several hurdles. Who knows the facts? Are they competent to testify? Are they deliberately distorting facts for their own benefit? Are their memory and powers of observation faulty? Take a common type of situation, such as an automobile accident, and think it over for a minute. Two cars collide suddenly. Question: Who saw what, and who can testify on how fast each car was going; where was it on the road; why didn't the drivers see each other and avoid the crash; were there any traffic control signs; and what were the conditions of the roadway? You can get conflicting testimony from practically everyone present. The witness may be doing his best to tell the truth, but his powers of observation may be poor, and the passage of time from the event to testifying in court further degrades his ability to describe clearly what happened.

In considering a trade secret, the question of proof—who knows and can testify on key facts—is often far more critical than what the law states on the point. A common conflict is on the significance of a conversation—was it a disclosure, and was it confidential? A written disclosure labeled "confidential" or a written agreement that the disclosure is confidential can avoid much litigation.

In considering an idea as nebulous as a trade secret, it is often illuminating to go back to the beginning. Our relationship with others might be well exemplified by the *Bible, Revised Standard Version,* Matt. 7: 12, "So whatever you wish that men would do to you, do so to them; for this is the law and the prophets." That is the golden rule; it is a good start but perhaps a little vague to set before a court. Let us try again with Exod. 20: 15, "You shall not steal;" verse 16: "You shall not bear false witness against your neighbor;" verse 17: "You shall not covet your neighbor's house, you shall not covet your neighbor's wife, or his manservant, or his maidservant, or his ox, or his ass, or anything that is your neighbor's."

The injunction against coveting thy neighbor's servant is quite pertinent as a prohibition against trying to hire a competitor's employee with the idea of getting trade secrets. Regarding verse 15, a new concept may arise. In stealing a trade secret, in one sense of the word, nothing is taken—i.e., the owner of the trade secret has all that he had before in a physical sense, and yet an idea can be more valuable than many concrete embodiments.

The courts are still pondering the question of whether a trade secret is "property." Consider the law of restitution—when should property be restored to its owner? If a trade secret is publicized, trying to restore secrecy would be worse than unscrambling an egg. A bill to revise the patent laws passed the Senate but expired on adjournment of the 94th Congress. Patent reform bills of varying scope have been in Congress for many years. Changes are needed. Another bill will undoubtedly be introduced shortly after the next Congress meets. The objective of the last bill was to increase the presumption of validity and provide for more fully disclosed inventions and other desirable objectives, but the bill seemed to throw out the baby with the bath water. A bill similar to the last one would increase the cost of patents markedly, and with reexamination after publication, the inventor loses the traditional exchange of his invention for 17 years' limited protection and has his invention spread before all countries of the world before he is sure of getting a U. S. patent. This increases the risk of disclosure without protection. Hence, more so than ever, an inventor needs to consider carefully the possibility of protecting his ideas by keeping them as trade secrets. This can result in a major loss of new technology to the public.

The Patent Bar was opposed to the bill unless it was revised extensively. The opposition was not selfishly motivated since the bill would create a great demand for new patent attorneys and eliminate unemployment among those in practice.

Basically, a trade secret belongs to the originator—although he often sells it to his employer as a condition of employment—and the employee should not compromise it. The originator should not try to sell it twice. On the other hand, the skill of a profession is the property of the employee. We have an analytical chemist who knows how to analyze carbon, hydrogen, and nitrogen using a combustion furnace and routine analytical techniques. Now what are those routine analytical techniques? There are many tricks to the trade and many different sources of error. Some of these are such that we learn them in elementary chemistry in college. Others may be known only in a few analytical laboratories.

A trade secret may even be well known. For instance, we have several methods of refining petroleum to make gasoline. The question is exactly which method and which conditions should be used for a particular feedstock. This can be an important trade secret even though the general aspects of refining are well known. As a good example of what can be a real trade secret and yet generally very well known, consider the combination to a safe. Anyone skilled in locks could look at a particular safe and know the general series of numbers used. However, whereas that ability could be said to be well within the skill of the profession, the exact numbers are a secret. If any employee were to disclose the combination to a safe to someone not authorized, he would be compromising a trade secret. Even if the employee is fired, we would agree that compromising the security of the safe is reprehensible. Of course, I might add that the safe should be reset using new numbers if any employee leaves.

Differentiating a trade secret from the skill of an individual can be difficult. A top-notch football quarterback is ready to coach his teammates, but throwing a pass entails long hours of practice as well as native ability and other factors. It is a skill not readily taught or learned, and there is nothing really "secret" about it, yet the coordination into a team effort involves managerial skills as well as trade secrets—the signals on the team can be a classic trade secret that can be compromised. There are subtle shadings between skills and secrets.

In chemistry, how many times is the trade secret of the employer based on some specific information or specific fact unique to that employer, the details of which should not be compromised? Four separate interests must be considered in a trade secret situation:

(1) The owner of a trade secret—often a former employer

(2) The recipient of a trade secret—often a new employer

(3) The conduit of a trade secret—often an employee who may be switching employment with his duties, obligations, and services to his masters to be separated, his skills to go with him, and trade secrets to remain behind

(4) The public—which is interested in promoting industrial growth and prosperity and full employment for all, including chemists.

Now consider who is to adjudicate these interests and what law will control—we have a federal government and 50 states. A case of major and recent importance is that of Kewanee Oil Co. vs. Bicron Corp.—particularly the decision of the Supreme Court on May 13, 1974 (2). In addition to the litigants, 21 different organizations filed briefs as *amicus curiae* (friends of the court). These groups wanted their views presented to the court because the decision in the case would be a precedent of concern to them. The American Chemical Society was among those presenting a brief.

Mr. Justice Douglas and Mr. Justice Brennan dissented with an opinion. Mr. Justice Marshall concurred with an opinion. Basically the decision reversed the Sixth Circuit Court of Appeals and held that the Ohio trade secret law is not preempted by the federal patent law. Much of the reasoning and comments are of present interest. In part the decision reads (footnotes omitted):

We granted certiorari to resolve a question on which there is a conflict in Courts of Appeals: whether state trade secret protection is pre-empted by operation of the federal patent law. In the instant case the Sixth Circuit Court of Appeals held that there was pre-emption. The

Second, Fourth, and Ninth Circuit Court of Appeals have reached the opposite conclusion.

Harshaw Chemical Co., a division of Kewanee, was able to grow a 17-in. crystal for ionization detection, which no one else had done. Harshaw considered the processes involved to be a trade secret. Several former employees of Harshaw formed or later joined Bicron Co. These employees had signed at least one agreement with Harshaw not to disclose confidential information or trade secrets. Bicron was formed to produce crystals.

Harshaw sued in the U. S. District Court under the Ohio trade secret laws and was granted a permanent injunction against disclosure or use of 20 of the 40 claimed trade secrets until such time as the trade secrets had been released to the public or obtained from authorized sources. The Sixth Circuit Court of Appeals reversed because:

Ohio could not grant monopoly protection to processes and manufacturing techniques that were appropriate subjects for consideration under 35 U.S.C. § 101 for a federal patent.

The Supreme Court reversed the decision of the Court of Appeals, holding that "Ohio's law of trade secrets is not preempted by the patent laws of the United States," and further held (footnotes omitted):

Ohio has adopted the widely relied-upon definition of a trade secret found at 4 Restatement of Torts § 757, comment b (1939). (B. F. Goodrich Co. v. Wohlgemuth, 117 Ohio App. 493, 498 (Ct. App. 1963); W. R. Grace & Co. v. Hargadine, 392 F.2d 9, 14 (CA6 1968). According to the Restatement at 5:

"(a) trade secret may consist of any formula, pattern, device or compilation of information which is used in one's business, and which gives him an opportunity to obtain an advantage over competitors who do not know or use it. It may be a formula for a chemical compound, a process of manufacturing, treating or preserving materials, a pattern for a machine or other device, or a list of customers."

The subject of a trade secret must be secret, and must not be of public knowledge or of a general knowledge in the trade or business. B. F. Goodrich Co. v. Wohlgemuth, supra, 117 Ohio App., at 499. National Tube Co. v. Eastern Tube Co., 3 Ohio C.C.R. (n.s.) 459, 462 (Cir. Ct. 1902), aff'd, 69 Ohio St. 560, 70 N.E. 1127 (1903). This necessary element of secrecy is not lost, however, if the holder of the trade secret reveals the trade secret to another "in confidence, and under an implied obligation not to use or disclose it." Cincinnati Bell Foundry Co. v. Dodds, 10 Ohio Dec. Rep. 154, 156, 19 Weekly L. Bull. 84 (Super. Ct. 1887). These other may include those of the holder's "employes [sic] to whom it is necessary to confide it, in order to apply it to the uses for which it is intended." National Tube Co. v. Eastern Tube Co., supra. Often the recipient of confidential knowledge of the subject of a trade secret is a licensee of its holder. See Lear, Inc. v. Adkins, 395 U.S. 653 (1969).

The protection accorded the trade secret holder is against the disclosure or unauthorized use of the trade secret by those to whom the secret has been confided under the express or implied restriction of nondisclosure or nonuse. The law also protects the holder of a trade secret against disclosure or use when the knowledge is gained, not by the owner's volition, but by some "improper means." 4 Restatement of Torts, § 757(a), which may include theft, wiretapping, or even aerial reconnaissance. A trade secret, however, does not offer protection against discovery by fair and honest means, such as by independent invention, accidental disclosure, or by so-called reverse engineering, that is by starting with the known product and working backward to divine the process which aided in its development or manufacture.

Novelty, in the patent law sense, is not required for a trade secret. W. R. Grace & Co. v. Hargadine, supra, 392 F.2d, at 14. "Quite clearly discovery is something less than invention." A. O. Smith Corp. v. Petroleum Iron Works Co., 73 F.2d 531, 538 (CA6 1934), modified to increase scope of injunction, 74 F.2d 934 (1935). However, some novelty will be required if merely because that which does not possess novelty is usually known; secrecy, in the context of trade secrets, thus implies at least minimal novelty. . . .

The only limitation on the States is that in regulating the area of patents and copyrights they do not conflict with the operation of the laws in this area passed by Congress. . . .

The Supreme Court considered the objective of the Ohio trade secret laws and considered that their laws were not at odds with the patent statutes. Neither removes matter from the public domain. If trade secrets were to apply only to non-patentable subject matter, an innovator would be at great risk in evaluating patentability. The court records on holding patents invalid clearly show that many inventors have invalid patents. To ask for a judgment on whether to seek protection as a trade secret or a patent puts too heavy a burden on the innovator. Quoting further:

The maintenance of standards of commercial ethics and the encouragement of invention are broadly stated policies behind trade secret law. "The necessity of good faith and honest, fair dealing, is the very life and spirit of the commercial world.". . .

Trade secret law provides far weaker protection in many respects than the patent law. While trade secret law does not forbid the discovery of the trade secret by fair and honest means, e.g., independent creation and reverse engineering, patent law operates "against the world," forbidding any use of the invention for whatever purpose for a significant length of time. The holder of a trade secret also takes a substantial risk that the secret will be passed on to his competitors, by theft or by breach of a confidential relationship, in a manner not easily susceptible to discovery or proof. Painton & Co. v. Bourns, Inc., supra, 442 F.2d, at 224. Where patent law acts as a barrier, trade secret law functions relatively as a sieve. The possibility that an inventor who believes his invention meets the standards of patentability will sit back, rely on trade secret law, and after one year of use forfeit any right to patent protection, 35 U.S.C. § 102(b), is remote indeed. Nor does society face much risk that scientific or technological progress will be impeded from the rare inventor with a patentable invention who chooses trade secret protection over patent protection. The ripeness of time concept of invention, developed from the study of the many independent multiple discoveries in history, predicts that if a particular individual had not made a particular discovery others would have, and in probably a relatively short period of time. If something is to be discovered at all, very likely it will be discovered by more than one person. . . .

We conclude that the extension of trade secret protection to clearly patentable inventions does not conflict with the patent policy of disclosure. Perhaps because trade secret law does not produce any positive effects in the area of clearly patentable inventions, as opposed to the beneficial effects resulting from trade secret protection in the areas of the doubtfully patentable and the clearly unpatentable inventions, it has been suggested that partial pre-emption may be appropriate, and that courts should refuse to apply trade secret protection to inventions which the holder should have patented, and which would have been, thereby, disclosed. However, since there is no real possibility that trade secret law will conflict with the federal policy favoring disclosure of clearly patentable inventions, partial pre-emption is inappropriate. Partial pre-emption, furthermore, could well create serious problems for state courts in the administration of trade secret law. As a preliminary matter in trade secret actions, state courts would be obliged to distinguish between what a reasonable inventor would and would not correctly consider to be clearly patentable, with the holder of the trade secret arguing that the invention was not patentable and the misappropriator of the trade secret arguing its undoubted novelty, utility, and non-obviousness. Federal courts have a difficult enough time trying to determine whether an invention, narrowed by the patent application procedure and fixed in the specifications which describe the invention for which the patent has been granted, is patentable. Although state courts in some circumstances must join federal courts in judging whether an issued patent is valid, Lear, Inc. v. Adkins, supra, it would be undesirable to impose the almost impossible burden on state courts to determine the patentability—in fact and in the mind of a reasonable inventor of a discovery which has not been patented and remains entirely uncircumscribed by expert analysis in the administration process. Neither complete nor partial pre-emption of state trade secret law is justified. . . .

Mr. Justice Marshall, concurring in the result:

Unlike the Court, I do not believe that the possibility that an inventor with a patentable invention will rely on state trade secret law rather than apply for a patent is "remote indeed." Ante, at 19. State trade secret law provides substantial protection to the inventor who intends to use or sell the invention himself rather than license it to others, protection which in its unlimited duration is clearly superior to the 17-year monopoly afforded by the patent laws. . . .

I conclude that there is "neither such actual conflict between the two schemes of regulation that both cannot stand in the same area, nor evidence of a congressional design to preempt the field." Florida Lime & Avocado Growers v. Paul 373 U.S. 132, 141 (1963). I therefore concur in the result reached by the majority of the Court.

This decision in the Kewanee case merits study in full, and but for the limit of space would be set forth more fully here. Neverthless, certain points can be emphasized. The source of patent protection is granted in the Constitution, Article I, Section 8, "The Congress shall have power . . . to promote the progress of science and useful arts by securing for limited times to authors and inventors the exclusive right to their respective writings and discoveries." Title 35 of the U. S. Code and other enactments of Congress are the controlling law. Congress can and does amend and change the law. Prior to the Constitution some states had their own patent laws.

The federal judiciary handles litigation on patents nearly exclusively. Theoretically, the same law governs all the judicial circuits. Where there is a conflict between circuits, as in the Kewanee case, sooner or later the U.S. Supreme Court can establish the controlling interpretation. Perhaps more in theory than in practice, the same decision should be obtainable in any U.S. district court.

In sharp contrast, in trade secrets each state has its own law. Historically, this law is derived primarily from judicial interpretation of the English common law. Some states use the Code Napoleon or Spanish law as a background. As set forth in Stamicarbon N.V. vs. American Cyanamid Co. (*3*), no fewer than 20 states during the past nine years have enacted statues making appropriation or unauthorized disclosure of trade secrets a crime. The decision lists the states. One commendable statute is that in New Jersey (*4*).

As a tort, or civil wrong against person or property, any legal action must be taken where the wrongdoer can be served and brought to court. Infringement of a patent is such a tort. As a crime, extradition between the states and from foreign countries greatly simplifies getting the wrongdoer into an appropriate court. Misuse of a trade secret by state statutes can be such a crime. The complexity of the law, with each state having its own laws, frequently far from uniform, can be illustrated by reference to Milgrim, "Trade Secrets" (*5*), which states, "practically all jurisdictions have recognized that a trade secret is property." It then cites cases from 30 states and 10 federal circuit courts of appeal. Milgrim's "Trade Secrets" (*5*) is an excellent text on the subject and far more detailed than we can be here.

Perhaps its author is trying to tell us something about the rate of change in trade secret law by publishing this treatise in loose-leaf form with yearly supplements. A U.S. patent has a fixed term of 17 years and extends protection to the United States only—although its disclosures are worldwide. Foreign patents can be obtained separately in each country and may be quite costly. The scope of protection and subject matter which is patentable varies tremendously. There is action towards patents common to several countries, but the cost is apt to remain high, and many years will be required to determine their effectiveness.

A trade secret has no geographical boundaries and no fixed lifetime. It expires when the subject matter becomes "generally known," so that it is no longer a secret. Comments on the point are too lengthy to explore here. Much of the law devolves from the centuries' old relationship of master and servant. If the master disclosed a trade secret to a servant, in confidence, the parties were of equal size. With modern corporations and individual employees the sizes are disproportionate. Still, however, a secret or confidential information can pass from one to the other. A corporation may spread a confidence among as many as have a reasonable need to know without the secret's losing its confidential status, but if spread too widely, the secret aspect may be lost.

The owner of a trade secret may at times protect it by an oral understanding, or the circumstances surrounding disclosure to those who need to know as a requirement for their tasks will speak for itself. The better practice, however, requires a written document. This may be in broad terms—e.g., each time the employee acquires new information, a new agreement need not be drafted. Substantial changes in relationships require updating of the agreement.

If the relationships between the previous employer, who owns trade secrets, the departing employee, and his new employer degenerate to the position that court action appears imminent, that action may well be taken in a state court or a federal court applying state law, and a jury may make the decision. A jury sometimes gives specific answers to specific interrogatories but usually delivers a verdict on the evidence without explanation. It is difficult, therefore, to predict what a jury will hold. It is also difficult to reason from the verdict in one case to another.

All that a jury knows about a situation is what various witnesses testify. When different witnesses give different stories, the jury must choose which to believe. With a jury, it is not uncommon for the sympathy to lie with the little man—the single employee who claims he has been done wrong.

No general rule can cover all cases because if a rule is set forth, the means of evading it are built in. As good a rule as any is the balance of equities—i.e., can each party truly say that if he were on the other side, he would feel justice had been done? One of our eminent Justices once said that he had trouble defining pornography, but he could certainly recognize it. Perhaps ethics in trade secrets fit the same class.

Consider the background in which the jury operates and the equitable position of the parties:

(1) The employee has deliberately stolen and delivered secret information while still keeping his old job as a cover.

(2) The employee and his associates have in effect formed a new company to profit by the stolen secrets.

(3) The employee has been induced to quit by the new employer.

(4) The employee quit in hope of getting a new and better job but without an offer in his pocket.

(5) The employee has taken deliberate action to be fired so as to be available for new employment, with or without a new job in his pocket.

(6) The employee has been fired for independent reasons.

(7) The employee has been given treatment deliberately designed to cause him to quit.

(8) The employee's job has been pulled out from under him because of reduction in force, or his former employer is disbanded or moving to an unacceptable location.

It is easy to see that the employee may by no fault of his own be forced to look for a new job. Any of these factors, if proved, could sway the equities and the sympathy and holding of the court or jury. Obviously, a jury is more sympathetic to an employee who is thrown out than to one who is trying to doublecross his employer. These imponderables and the relationship of a trade secret to the employee's chances of obtaining a job elsewhere influence the likelihood of an injunction or fine being imposed for compromising a trade secret.

In time of war, spies are shot and traitors are hung; they are dealing in a special class of trade secrets. A tremendous amount of publicity was generated in the case of compromising certain nuclear secrets, and the Rosenbergs were executed. Those who compromise industrial secrets are usually treated more gently. Also to be considered are whether compromising of the trade secret was the prime factor in a new job offer or whether it was incidental or accidental. The time lapse between the depature of an employee and his compromising a trade secret is an intangible factor that looks towards intent. The number of people who know the secret, the part that the employee had in developing the secret knowledge, and its relationship with his field of employment all are factors. Consider a programmer who takes a reel of tape with a key program to a new employer. That action would be reprehensible even if he wrote the program, but if he takes his skill in writing programs and develops completely new programs, that is not only acceptable, but it is customary. If he remembers and takes the concept only with him, then the situation can get sticky, and the comparative rectitude of the parties comes into the situation.

One example of resolution of possible conflicts of interests is Patent Office Rule 341(g) which provides that no person who has served as a patent examiner is permitted to prosecute or aid in any application pending in his examining group during his service therein, and he is not permitted to prosecute within the group for a period of two years after his leaving. 18 USC 207 prohibits a former employee of the government from acting in any case in which he participated as an employee, forever, and for a period of one year he must not appear before an agency which was under his responsibility, with a penalty of $10,000 and/or two years' imprisonment. Anyone who is particularly interested in the point should read the fine print.

The Code of Professional Responsibility of the American Bar Association, Disciplinary Rule 2-108, provides in part that a lawyer shall not be a party to or participate in a partnership or employment agreement with another lawyer that restricts the right of a lawyer to practice law after the termination of a relationship created by the agreement, except as a condition to payment of retirement benefits. DR 4-101 provides for the protection of confidences and secrets of a client presumably forever unless disclosure is necessary under conditions set forth in the fine print, as, for example, to collect his fee. Interestingly, DR 2-106 (B) (2) provides that the determination of a reasonable fee includes "the likelihood, if apparent to the client, that the acceptance of the particular employment will preclude other employment by the lawyer." A lawyer cannot serve both sides in a dispute, and taking one client may later bar lucrative employment.

Similarly, some employment agreements in technical fields provide that a departing employee will not accept employment at variance with the interest of his previous employer for a reasonable fee for a limited period of time—say six months to a year on payment of his salary. Consider such secrecy protection in connection with the chapter on employment agreements.

Some readers would like to get clear and absolute rules on what a chemist can consider a trade secret without breaking the law. An analogy can be used. The speed limit in a school zone is 15 mph. At 2 a.m. on a clear night, a much greater speed would be acceptable than during school recess when a large number of children are in the immediate vicinity. It also depends on who is measuring the speed and how accurate the measurement is. A speed of 15 mph, as long as the car is not a hazard to children, is the safe rule, but many drivers move faster without being called to account for this action in court. The fact that many others have not been charged is not an adequate excuse for the driver who gets the speeding ticket.

With trade secrets, a great many more individuals bend the law a bit than are called to account for their actions. The owner of a trade secret may feel that his cause is just but that a victory would not be worth the effort. Also some employers feel that if an employee will steal for them, he will steal from them, and they do not want an employee with too flexible a conscience. The exact lines in trade secrets are very vague and the subject of many differences of opinions. There are hundreds of reported cases. Thousands of situations have been considered and never brought to court. To cite a few examples, in Wilkes et al. vs.

The Pioneer American (6), the district court held that trade secrets could consist of conception, formulation, and development of a method of selling life insurance. The court held that the fact that part, or even eventually all, of the components of a trade secret are matters of public knowledge does not prohibit a claim of trade secret. An injunction was granted. It was pointed out that facts of great value may live long in the public domain unnoticed. The protection is not on the secret alone but against the breach of faith and reprehensible means of learning that secret. It is interesting to note that technical and commercial trade secrets serve as precedents for each other in court cases on breach of a confidential relationship.

In Canada, Carnaghan Insurances Ltd. vs. Lundy (7) involved a restrictive covenant; the employee was valuable and held trade secrets in the insurance field. The court held that a covenant not to compete for three years was reasonable but that the plaintiff's business was in Saint John's (the agreement covered the Province of New Brunswick). Under the circumstances, the area was too broad, the court declined to rewrite the area, and the plaintiff lost.

Sometimes a court will limit the time, area, or scope to hold a secrecy agreement valid. More often, the agreement is held not valid because it is too broad, and the agreement fails completely, even if a more limited agreement clearly would have been proper. The court declines to rewrite the agreement.

In K-2 Ski Company vs. Head Ski Company and William Crocker (8) the court approved a two-year injunction on certain phases of ski manufacturing and a one-year injunction on another phase. The question of damages was remanded to the lower court. Here Crocker had learned all he knew about ski manufacturing while working for K-2 from May 1967 to Feb. 1970. His responsibilties were reduced, he became dissatisfied, so he contacted Head and was offered a job. He then quit K-2 without disclosing the name of his new employer. The opinion does not mention any secrecy agreements but does find that security at the K-2 plant was not tight. This is an example of an employee leaving for what amounted to an undesirable change in working conditions.

A. H. Emery Co. vs. Marcan (9) involves both a patent and trade secret situation. The patent was invalid. The former employer prevailed on the trade secret aspects. A confidential relationship survived a termination of employment even though one of the former employees was dismissed for other reasons before the group of former employees started their activities.

The express labelling of drawings as "confidential" or "secret" was not necessary if they were, in fact, confidential and were taken. Conversely other cases have held that labelling non-secret data as "secret" does not make it secret.

In American Cyanamid Co. vs. Fox et al. (*10*) the employee stole trade secrets, including microorganisms for the production of antibiotics, and sold them internationally. Fox and his associates had compromised many millions of dollars worth of trade secrets, but retrieving the stolen secrets was impossible. A judgment for damages is futile if the defendant has no funds. The record in the case consisted of 6,984 pages.

In Sperry-Rand Corp. vs. Rothlein (*11*), Rothlein felt he did not get a fair break on promotions and decided to leave Sperry-Rand and start his own competitive semiconductor company. He picked up 21 of 35 employees from Sperry-Rand for the new company. Sperry-Rand was held entitled to damages. The length of the proceedings, about five and one-half years, did much to make an injunction futile. The fact that the ex-employees developed the process gave them no greater right to use it competitively than any other employee.

In Sperry-Rand Corp (*12*) vs. Electronics Concepts, Inc., et al. the ex-employees had taken valuable data on radar antennas. One had learned what he knew at Sperry-Rand and went into competition with Sperry-Rand—even competing on a government bid with stolen knowledge of the exact amount of the Sperry-Rand bid. Both damages and an injunction were awarded to Sperry. These ex-employees willfully and deliberately and with full knowledge of the unlawfulness of their acts took both technical and financial data.

Prognosis

To speculate on what the law will be in the near future is risky at best and in trade secret law, perhaps more so. Certain trends seem to be present. There is a bill in Congress on a Federal Unfair Competition Act, presently S.31 in the 95th Congress, which will probably be before Congress for some time in substance under different numbers for each Congress.

Some of the states are adopting a Uniform Deceptive Trade Practices Act and a Uniform Trade Secrets Act. As more states adopt a Uniform Trade Secret Act, the laws in the various states tend to become more uniform and predictable. There is a trend towards a more uniform policy and punishment of individuals and organizations that participate in the outright theft of trade secrets. Hopefully this trend will continue.

On the other hand, there is increasing tendency towards permitting an employee to realize his potential both professionally and financially by availing himself of employment opportunities. An individual with scarce skills and unique knowledge who is considering employment opportunities might think about whether the prospective job has a disproportionate income purportedly attached. If an employer makes a job offer that is too far out of line with the going rate, this raises the question of whether he is really trying to encourage theft of trade secrets, whether, once he picks the brain of his scientist, the job will evaporate, or whether the new employer seeks to use the legitimate skills of the new employee towards a long term objective.

From the ethical standpoint, a patentable invention or a trade secret is like a piece of real estate: it can be sold once, but then the original owner has no right to sell it again. On the other hand, we might compare an inventor with spectacular skill with the owner of an apartment house who can rent out an apartment and collect rent each month. Similarly, the scientist with unique skills can rent out those skills as an employee of a single entity or as a consultant to several, but he should be sure that he is renting skills and not purloining trade secrets.

The heart of misuse of trade secrets is really the breach of confidence. Confidentially acquired information in any field needs to be protected. It is to be regretted that clear guidelines cannot be established, but the tremendous number of court cases shows that differences of opinion intertwined with problems of proof are constantly arising. Certain treatises, texts, and references are appended. Some of these, like Milgrim's "Trade Secrets" (5), have thousands of citations to court cases. There are many additional texts as well as law review articles on various aspects of trade secrets. The standard legal digests will locate cases which are pertinent to particular situations. Regretfully, in the space available here, it is possible to touch only a few highlights and to give a few clues as to further research on the subject.

In purely chemical cases, the evidence as to what was taken, proof that it was a trade secret, and that it was not known to competitors in academia or the literature becomes too complex to set forth in a short summary. Generally, a reputable employer will avoid trying to buy a trade secret with a new employee. An employee from one company can be used by his new employer in an operation that does not compromise trade secrets. The best rule is whether it seems fair to all concerned. When it gets close to the edges, watch out. Litigation often is such that nobody really wins; it is just a question of how badly everybody loses.

Summary

(1) Patent Law is statutory with essentially exclusive federal jurisdiction.

(2) Patents have a life of 17 years.

(3) Trade secret protection is essentially a matter of state law with about 52 jurisdictions, not necessarily consistent.

(4) Trade secrets are secret as long as they can be kept secret. There is no theoretical or statutory limit on their life.

(5) Much of the State Trade Secret Law is judicially written based on old common law precedents. Some states have modernized it with state statutes.

(6) Whenever the nature of an invention is such that it can be exploited in secret, thought should be given to trade secret protection rather than patent protection.

(7) If a new patent act changes the patent statutes, its impact on trade secrets should be considered as to length of protection, costs, and effectiveness of protection.

(8) An employee should respect the confidences of his employer, even long after employment has ended.

(9) The former employer should expect an ex-employee to use the skill of his calling in a new employment.

(10) No matter how skillfully are confidentiality or secrecy agreements written, high ethical standards on both sides are needed. Larceny runs deep in the human soul, and if something has value, someone will always try to steal it.

Literature Cited

1. Commonwealth vs. O'Neal, 339 N.E.2d 676.
2. Kewanee Oil Co. vs. Bicron Corp., 416 U.S. 470, 94 SC 1879, 181 USPQ 673 (1974).
3. Stamicarbon N. V. vs. American Cyanamid Co., 506 F.2d 532 (C.A.2., 1974) at 540.
4. N.J. Stat. Anno. 2A: 119-5.1 et seq. Compare N.Y. Sec. 1296 and penal code 155:00 and following.
5. Milgrim, R. M., "Trade Secrets," Matthew Bender, New York, 1967, with annual supplements. Note particularly Chap. 5 on employer–employee relationships.
6. Wilkes et al. vs. The Pioneer American, 383 F. Supp., 185 USPQ 95 (DC SC 1974).
7. Carnaghan Insurance Ltd. vs. Lundy, 20 *Can. Pat. Rep.* **2d**, 184.
8. K-2 Ski Company vs. Head Ski Co. and William Crocker, 506 F.2d 471 (CA9, 1974).
9. A. H. Emery Co. vs. Marcan Products Corp. et al., 268 F. Supp. 289, 153 USPQ 337 (DC S NY 1967) affirmed 389 F.2d 11, 156 USPQ 529 (CA2, 1968), Cert denied 159 USPQ 799.
10. American Cyanamid Co. vs. Fox et al., 140 USPQ 199 (NY Sup Ct. NY Cty-1964).
11. Sperry-Rand Corp. vs. Rothlein et al., 241 F.Supp 549, 143 USPQ 172 (D.C. Conn. 1964), 288 F.2d 245 (CA2, 1961).
12. Sperry-Rand Corp. vs. Electronics Concepts, Inc. et al., 325 F.Supp. 1209 170 USPQ 410 (D.C. E.D. Va., 1970), modified as to damages 447 F.2d 1387, 171 USPQ 775 (CA4, 1971), Cert denied 173 USPQ 193 and 175 USPQ 385.

Bibliography

Abbott Labs. vs. Norse Chemical Corp., 152 U.S.P.Q. 640, 649 (1967).

Arnold, T., "Trade Secrets, Legislative Objectives and Proposals," *IDEA* (1965) **9**, 161.

Arnold, Jr., T. H., "Are You Locked into Your Job by What You Know?" *Chem. Eng. J.* (1966) **141.**

Blake, H. M., "Employee Agreements Not to Compete," *Harv. Law Rev.* (Feb. 1960) **73**, 625.
Biesterfeld, C. H., "Patent Law," 2nd ed., Wiley, New York, 1949.
Bowen, W., "Who Owns What's in Your Head?" *Fortune* (July 1964) **175.**
Brooks, J., "Annals of Business—One Free Bite," *New Yorker* (Jan. 11, 1963) **37.**
Carter Products, Inc. vs. Colgate Palmolive Co., 130 F. Supp 557, 572, 104 U.S.P.Q. 314 (D. Md. 1955) 108 U.S.P.Q. 383 (C.A. 4) 352 U.S. 843 (1956).
Choate, R. A., "Patent Law," West Publishing Co., St. Paul, Minn., 1973.
Deller, A. W., "Deller's Walker on Patents," 2nd Ed., Baker, Voorhis & Co., Inc., Mt. Kisco, N.Y., 1964 (yearly supplements).
Eckstrom, L. J., "Licensing in Foreign and Domestic Operations," Revised 3rd ed., Clark Boardman, Co., Ltd., New York, 1972 (loose leaf supplements).
E. I. duPont de Nemours & Co., Inc. vs. American Potash and Chemical Corp., 200 A.2d 428 (Del. Ch. 1964).
Ellis, R., "Trade Secrets," Baker, Voorhis, and Co., Inc., New York, 1953.
"Employee Patent and Secrecy Agreements," Studies in Personnel Policy, No. 199, Nat. Ind. Conf. Board, Inc., New York, 1965.
Gray, A. W., "Are You Free to Change Jobs?" *Chem. Eng.* (1962) **69**, 126.
J. Pat. Off. Soc., P.O. Box 2600, Arlington, Va., published monthly.
Klein, H. D., "The Technical Trade Secret Quadrangle: A Survey," *Northwest. Univ. Law Rev.* (1960) **55**, 437.
Leydig, C. F., "Protecting Trade Secrets," *The Bus. Lawyer* (1966) **21**, 325.
Liebhafsky, D. S., "Industrial Secrets and the Skilled Employee," *N.Y. Univ. Law Rev.* (1963) **38**, 324.
Melville, L. W., "Forms and Agreements on Intellectual Property and International Licensing," 2nd ed., Clark Boardman Co., Ltd., New York (loose leaf) 1972.
Milgrim, R. M., "Protecting and Profiting from Trade Secrets," Practicing Law Institute, New York, 1975.
Munster, Jr., J. H., Smith, J. C., "Savants, Sandwiches and Space Suits," *Science* (1964) **145**, 1276, and "The Care and Feeding of Intellectual Property," ibid. (1965) **148**, 739.
Sutton, J. P., "Trade Secrets Legislation," *IDEA* (1965) **9**, 587.
"Trade Secrets—Ethics and the Law," American Chemical Society, Washington, D.C. (1968).
"Trade Secrets, the Technical Man in Legal Land," *Chem. Eng. News* (1965) **43**, 80.
Turner, A. E., "The Law of Trade Secrets," Sweet and Maxwell, Ltd., London, 1962.
Wade, W., "Industrial Espionage and Misuse of Trade Secrets," Advance House Publishers, Ardmore, Pa., 1964.

RECEIVED October 29, 1976.

5

Employment Contracts

JOHN P. SUTTON

3000 Ferry Building, San Francisco, Calif. 94111

The common provisions in current employment agreements relate to the disclosure of patentable inventions, cooperation in obtaining patents, assignment of inventions to employers, and protection of trade secrets. Wide variations exist in the presentation of these provisions in the agreements. As instruments of corporate personnel policy, these agreements almost universally favor the employer. The author suggests that the American Chemical Society gather data through efficiently conducted surveys and seek to make employment agreements bilateral, encouraging employers to abide by the ACS Guidelines for Employers.

The majority of inventions made in the United States today are made by employees of corporations. These employed inventors are required to sign written employment agreements in 98% of the cases (*1*). Why are such contracts so popular? The common law (i.e., the law that applies if there is no written agreement) governing inventions by employees appears to be straightforward. Professor Stedman puts it this way:

> Where the facts show that the employee was specifically hired to make inventions, the inventions that result belong to the employer, and the employee is required to assign them to him. This doctrine applies, however, only to those inventions that fall within the field for which he was actually hired and not to inventions he may make in other areas. . . . if the employee engages in inventive activity that is entirely independent of his job, e.g., work done at home in areas not related to his employment and not involving the use of his employer's facilities or time, the inventions that result belong entirely to the employee just as though he were unemployed (*2*).

Disputes arising over whether an employee is hired to invent in the scope of his work assignment cause the problems. Employment agreements, which vary widely from corporation to corporation, are placed before a newly hired employee to avoid these disputes. They have common features yet enormous differences. Companies usually make a flat payment of $100 to $200 to employed inventors upon filing of an application, issuance of a patent (*2, 3*), or both. This provides the incentive for an employee to disclose his invention and thus fulfill the purpose of the patent system. Recently, however, I conducted a survey of employed inventors in California (*1*), which revealed that although 57% of corporate employers had sales over one million dollars annually, a majority of inventors responding (54%) indicated they received $1 or nothing in direct compensation for their inventions.

This finding contradicts Neumeyer and O'Meara (*2, 3*) who found that most large companies, 60% of them in O'Meara's evaluation, gave monetary rewards to employed inventors. Having failed to receive adequate support in another survey of employers in order to clear up the discrepancy, I have proposed that the American Chemical Society (ACS) sponsor such a survey since they possess the resources to complete it successfully. Based on the small amount of response (20%) I received in my survey, however, and on 10 other employment agreements given to me by the ACS Council Committee on Professional Relations, I have found certain common features in these agreements.

Only one clause was common to every one of the agreements I studied in detail: a duty of cooperation by the employee with respect to patenting of inventions. The second most common feature is a duty on the part of the employee to disclose inventions to his employer. Another common feature was a duty to refrain from disclosing trade secrets belonging to the employer, and a duty on the part of the employee to assign his inventions to the employer. It is fair to state that these are the usual provisions.

Besides these four similarities, however, other correlations do not exist to any great extent. Most of the agreements have a provision that states that the agreement is binding upon the successors and assigns of the employee, the employer, or both. Usually the agreement is binding on the employee's heirs and assigns and not on the employer's. This raises an important aspect of employment agreements. When one thinks of an agreement, he thinks of promises made by two parties as they relate to each other. However, there is a kind of contract—a unilateral contract —which binds only one party. In my modest survey I did not find a single employer who promised to do anything for the employee. One of them purported to have presented a bilateral agreement, but there was no provision for the employer to sign. Such an agreement would not be binding, ordinarily, on any party who does not sign a contract. On the other hand, the majority of contracts provided to me by the Council Committee on Professional Relations were bilateral and included a place for the employer to sign, binding him to the terms of the agreement.

Whether the agreement was bilateral or unilateral, none of the employers promised his employee anything other than continued employment for an unspecified time. In only one agreement, from a small southern manufacturing company, was there any provision for compensation, and this agreement simply said that the employer may compensate the employee for inventions. None of the agreements in either category made a promise to pay even the traditional $100 award. These award programs appear to be governed entirely by corporate policy and not by contract. This means that the corporation may change the policy at will without consulting the employee. The employee, on the other hand, cannot vary any of the terms of the employment agreement.

Several of the agreements that were studied had express provisions requiring an employee not to engage in any outside employment within the area of interest of the employer. Perhaps a moonlighting policeman or fireman could work for these companies as a chemist, but a chemist could not use any of his skills and technical training in moonlighting jobs when employed by these companies—even if he works on projects unrelated in product lines of his first employer. It is enough in these agreements for the first employer to merely have an interest in an area, whether or not he actually does business in that area.

A number of the contracts went further than simply requiring trade secrets and inventions to be protected. They included a provision that a property right existed in the records, drawings, and other materials used in the employment, so that the employee would be bound whether or not the inventions or trade secrets contained in the records were protectable.

In none of the agreements received as a result of my survey was there a noncompetition clause. This is a provision by which an employee promises not to compete with his employer for a specified period of time. In three of the agreements provided to me from ACS, however, there were non-competition provisions. The most onerous one was a promise not to compete for one year plus the duration of any litigation that might arise concerning the subject matter of the agreement. This means that if the employee were sued within a year after his departure, he would not be able to compete until there had been a final determination from which no appeal could be made. This could easily be five years in all, given the crowded dockets of trial and appellate courts.

While a promise not to compete was rare, it was commonplace to have a provision that inventions completed within six months or a year after termination of employment would have to be reassigned to the former employer. If this provision were rigorously enforced, a new employer would be reluctant to assign a new employee to any area where he would be likely to make an invention within the first six months or a year. I do not believe such provisions are rigorously enforced.

A fairly common provision in the agreements was a requirement that the employee return all property to the employer upon termination of employment. Chemists are frequently pack rats, and it is difficult to determine precisely what belongs to the employer and what belongs to the employee. It is reasonable, however, to require that the employer's property be returned.

Another common provision is an opportunity for the employee to exclude inventions made in prior employment. Some space is provided for the employee to list on the agreement inventions which he has previously made and which belong to prior employers.

The agreements which were studied in detail were surprisingly free of the boiler-plate legalese which one often associates with license agreements and other contracts. Several included provisions that the governing law would be that of a particular state, that provisions which were unenforceable would be severed from the agreement so that the remaining provisions would be enforced, and, as mentioned previously, that the terms of the agreement would be binding on successors and assigns.

The lack of a non-competition clause in the agreements derived from the employer survey apparently reflect a concern that such a clause might not be enforceable. In California, for example, the Business and Professions Code §16,600 voids such clauses. This provision was recently upheld by U. S. Supreme Court (*4*). In other jurisdictions, there is a substantial risk that a clause prohibiting a chemist from engaging in his profession with another employer would be ruled unenforceable as an unconscionable contract of adhesion. It seems the trend of the law is to recognize the doctrine of contracts of adhesion as being unenforceable.

A contract of adhesion occurs when the terms are prepared entirely for the benefit of one of the parties, and the other party does not have sufficient bargaining power to alter the terms. Today the employment agreement is a contract of adhesion. Whether it is enforceable or not depends on whether it is unconscionable. Automobile warranties, insurance contracts, and other kinds of contracts have, in some instances, been ruled unconscionable contracts of adhesion. However, I know of no case dealing with an ordinary employment agreement although there is at least one case pending which raises this issue.

Not one of the agreements that I reviewed in detail provides for the employee to share in the benefits derived from his invention. Legislation is pending in the U.S. Congress and in California which would require the employer to share the benefits received from an invention by an employee, bearing in mind the position of the employee, his duties, the value of the invention, and the contribution of the employer. These laws are patterned after similar legislation in other countries. Nearly every industrialized country has legislation mandating extra compensation for employed inventors with the exception of countries whose law is derived from the English common law (U.K., Canada, U.S., and Australia) (5). This legislation is unlikely to pass in the foreseeable future.

Even though the study I have concluded does not show such agreements, employment contracts which provide that the employee will receive a specified percentage of any royalty income derived from licensing the invention do exist. This provision is fairly common in the aerospace industry but generally nowhere else. Should the invention have substantial value and be widely licensed, the inventor could derive substantially more than the usual payment of $100 to $200. In my first survey of California inventors, only 3% of the inventors received between $500 and $5000 for the invention, and not one received more than $5000. As previously indicated, 54% received $1 or nothing.

Some of the provisions in individual employment agreements were surprisingly one-sided in favor of the employer. One had an express provision that the agreement does not bind the company to pay any salary to the employee or to employ the employee for any period of time. Another agreement provided that the employee must pay attorney's fees and expenses and consent to a preliminary injunction in the event of litigation over a breach or threatened breach of any provision of the agreement. It seems grossly unfair to require the employee, who has relatively few resources compared with the employer, to pay these fees and expenses merely upon the existence of what the employer regards as a threatened breach of some term in the employment agreement.

Another employment agreement has an express provision that the employee must serve faithfully and to the best of his ability and to devote his entire time, energy, and skill to promote the corporate interests. It could be argued that many every day activities of an employee do not promote the corporate interest, such as going home at five o'clock even though an experiment is not completed.

One agreement, presumably intended to show the magnanimity of the employer, provides that it "will give consideration to the reassignment to the employee of any inventions . . . which it may find to be of no potential value to the company." There is no promise to do anything except consider. This attitude is the same as that expressed by Neumeyer (2) regarding award plans:

> Many of these have the character of a patriarchal, 18th-Century attitude toward the employee, a pat on the shoulder by the patron, who knows best.

My study of employment agreements gave broadly similar results as reported in 1965 by O'Meara, though a major distinction is the apparent lessening today of post-employment restrictions. That is, in 1965, 25%

of the agreements studied limited the employees activities after termination of employment, and my study showed very few such limitations.

Present-day employment agreements vary widely, even though they have four general features: they require assignment of inventions; they require non-disclosure of trade secrets; they require disclosure of inventions; and they require cooperation in prosecuting patents. The employment agreements of today do not bind the employer to compensate the employee for making an invention, and they do not recognize any substantial rights of the employee. As instruments of corporate personnel policy, they are oppressive and unfair, but universal. By contract, employers in the United States have effectively defeated the provision in the Universal Declaration of Human Rights adopted by the United Nations General Assembly on December 10, 1948. In Article 27(2) it states:

Everyone has the right to the protection of the moral and material interests resulting from any scientific, literary, or artistic production of which he is the author.

The American Chemical Society should seek to have employment agreements made bilateral, with the employer agreeing to abide by the Guidelines for Employers. I have never seen an employment agreement remotely similar to the provisions of the Guidelines for Employers. Individual employees will not be able to modify employment agreements because of their relatively weak bargaining power. If the agreements are to become fair and equitable to the interests of everyone, the balance between the employee and his employer will have to be readjusted, either by a profession-wide organized effort or by legislation to repair the imbalance.

Literature Cited

1. Sutton, J. P., "Compensation for Employed Inventors," *Chem. Technol.* (Feb. 1975) p. 86.
2. Neumeyer, F., "The Employed Inventor in the United States," MIT Press, 1971.
3. O'Meara, "Patent and Secrecy Agreements," National Industrial Conference Board No. 199, New York, 1965.
4. Merrill Lynch, Pierce, Fenner, and Smith, Inc. vs. Ware, 410 U.S. 908 (1973).
5. Neumeyer, F., "Systems to Stimulate Employee-Inventions in Europe," *NBS Special Publication 388,* U.S. Dept. of Commerce, Bureau of Standards, 1973.

RECEIVED September 17, 1976.

Discussion

Q. I was curious about your comments that the state of California has voided all non-competition agreements. Are there any circumstances under which the state of California would recognize such an agreement?

A. In the California statute, there are specific provisions that are exceptions to this. One is that a partnership that disbands can have a restriction so that there is no competition between the partners in a geographical area. The sale of a business can have a provision that there is no competition between the buyers and the sellers for some period of time. I didn't cover the exceptions—there are three or four of them—because they are quite specific and they don't apply to 99% of the chemists who would be here today, but there are exceptions.

Q. In past years the Supreme Court of Pennsylvania has held that an employee cannot be presented with a non-competition agreement to sign with the implied threat that if he doesn't, he will be fired. Such agreements are no longer enforceable. However, if a potential employee is presented such an agreement, and it constitutes one of the terms by which he is hired for the job, then the agreement can be enforced as long as there is consideration, and this has to be monetary in form.

A. Do you know the name of that case? I am not familiar with it.

Q. My attention has been drawn recently to an employment agreement which requires the prospective employee to sign an authorization by an employer to allow a consumer reporting agency to inquire into many personal aspects of his life. Doesn't this seem to go as a rather undue invasion of personal privacy to ask an individual to sign such an agreement? This is in fulfillment of Public Law No. 91-508.

A. I am not familiar with it. Nothing like that was brought into the survey I conducted or in any of the agreements that I studied. I haven't seen anything like an invasion of privacy at all. There could be some federal law that demands that, but I am not familiar with it.

RESPONSE FROM AUDIENCE: I believe that that is a response to the Privacy Act which now requires that if you do utilize such sources of information, there has to be a release from the individual involved.

Q. My question relates to this reassignment clause. Apparently the employee has to reassign his invention to the employer for $1 without knowing what the value is going to be. Do any courts have a rule about this?

A. You don't even have to have $1. The mere fact is that employment is sufficient in the majority of the cases. One of the four of the universal provisions I found was that you assign in advance, as a condition of employment, any inventions you make in the future. In other words, on day one you sign the agreement that whatever the invention value is and however the inventions are related to the business, they will be assigned. Now, some of them had a limitation that they had to be within the areas of interest to the corporate employer, but some did not.

Q. Do you have any suggestions as to bilateral agreements? What type of promises do employees have to make the employer for a bilateral agreement? How does an employee get a bilateral agreement binding on the employer?

A. I have all kinds of suggestions. The problem is that unless you are Nobel Laureate you are not going to get those provisions into the contract. It's bargaining power that gets fair contract provisions. Maybe if you have a Nobel Prize in the field you might be able to get a chauffeur and a limousine in your contract, or you could get other provisions which would be somewhere along the lines of the "Guidelines for Employers." Those in demand, like corporation presidents, can write their own tickets. Most employees cannot.

Q. What about a situation where an employee signs a contract where he promises to invent but does not intend to do so. Assume he does not invent. Does he breach his contract?

A. That is the way it used to be in the last century. In fact, chemists 100 years ago weren't as plentiful as they are today, and the chemist was hired by the job. Assume you've got this job requiring explosives, and the chemist was an expert in explosives. The chemist comes in to solve the problem and he says, "pay me so much to do this job and pay me so much in royalties on use of the explosive or if you license it to someone else." Usually the agreement would have some modest amount of living expense while the chemist was working on the project plus some return on the invention, sharing the benefits of the invention. If the invention were widely used, the inventor became rich, but those days are long past.

Q. So, you don't think a bilateral agreement is profitable?

A. Oh, I certainly do think it is profitable. Yes, I definitely think that it is profitable.

Q. Is that the way bilateral agreements work today?

A. That is the way it ought to be. I am telling you that the bargaining power between a chemist and his employer is so grossly disproportionate that he is not going to be able to write that kind of contract—fine chemist that he is.

Q. To the contrary, I think that employers would be glad to give to someone, a research person, an agreement in which the chemist promises to invent something of value in exchange for his being paid. The employment agreement of today, I think, serves the purpose very well. The employer promises to pay as long as the employee makes inventions and does his job. You think that that is not proper?

A. I think it is not factual. The chemist is hired to do research and solve problems whether they amount to inventions or not. If they do amount to inventions, it becomes a windfall for the employer. Inventions are, by definition, windfalls for someone, because they are creations where nothing existed before. I believe inventors, as indispensible creators, should share in the windfalls with the employer who puts up capital, poses the problem, and provides the environment.

6

Legalization of Employment Guidelines

A. C. NIXON

2140 Shattuck Ave., Berkeley, Calif. 94704

There is a pressing need for legislation in the area of professional employment guidelines. Employers make no commitment to their employees except to obey discrimination statutes. Any contracts that are made usually only deal with protection of trade secrets and assignment of patent rights. Professional employees are not covered under the Wage and Hours Act; therefore, they are not eligible for overtime pay. "Whistleblowers" must be afforded job protection so that they won't be inhibited by fear of dismissal from coming forth with information beneficial to the public welfare. The American Chemical Society and the Engineers Joint Council have guidelines they wish employers to follow but have no means of enforcing them. The author advocates making the core provisions of these guidelines into law and discusses the responsibilities of professional societies after such legislation.

Since time immemorial there have always been agreements between employers and employees. In the earliest days these were verbal, they followed custom, and there was generally no mechanism set up to protect the more vulnerable party—the employee. Of course, in many cases the employees had no rights whatsoever because the employees were owned by the employer. As time went on, codes of ethics were developed by different cultures which were applied to a greater or lesser degree to the relation between employer and employee. The Old Testament, for instance, spelled out some fairly concrete rules in this matter, and aggrieved employees had avenues of appeal through the Rabbinate or the Royal Court. During the Middle Ages the growth of guilds tended to stabilize employment by the semi-monopoly so created. The outgrowth of these guilds is the modern development of industrial unions. However, in this country the basic law has been inherited from English common law. In England the relationship between master and servant was based on custom which became codified into common law where there was an understanding of what the two parties were supposed to do for each other (in actual fact it appears that it was difficult for an employee to bring any legal sanction against employers who violated the code). The translation of the English common law into American law seems to have provided even less protection for the employee than exists in England.

Necessity for Guidelines for Employment

The employment conditions for most professions at the present time stem directly from the old English common law master–servant relationship. In other words, there is no commitment on the part of the employer to treat the employee in any particular way except as mandated by the laws against discrimination with respect to age, sex, racial origin, or religion. Also, individual contracts that are signed generally only require that the employee will not divulge his employer's trade secrets and will sign over to his employer any patentable inventions that he might make. Professionals are not even included under the Wages and Hours Act, so they are not eligible for pay for any overtime that they are required to work. Most professional employees can be fired for any reason whatsoever or for no reason at all.

This situation does not apply to academic employees who are tenured or to many government employees covered by Civil Service regulations. However, with the financial crunch that has affected academia and the complexity of the bureaucracy in government, individual employees rights are often lost sight of. Nonetheless, in theory at least, these two groups of employees have the freedom to speak out on public issues and to be relatively protected from mass layoffs. In fact, this is not entirely true. Even in cases of the tenured academic faculty, many have such complex and demanding ties to either industry or government that their freedom of expression, they feel, is severely curtailed. Numerous instances have come to light (I'm sure many more are hidden) in which government employees have found themselves transferred to less desirable or even non-existent jobs as a result of attempting to correct some inequity in public policy. And even if an individual employee is not thinking about speaking out on a matter of concern to the public, he should know how his performance is judged by his employer and that he is shielded from capricious actions on the part of his supervisor. Thus, if there are to be reductions-in-force, he should have some assurance that if his performance has been satisfactory, he will not be included in those that are fired as a result of the whim of some member of management.

From the standpoint of the public, however, the most important aspect of the master–servant relationship being applied to professional employees is that it is very inhibiting to the so-called "whistleblower." There have been many instances over the past several years where people have spoken out and suffered for it, and we know of many instances where it would have been greatly in the public interest to have knowledgeable professional employees come forward with information. Instances that come readily to mind are Fitzgerald and the C-5 aircraft situation (*1*), the three BART engineers, and more recently the three General Electric nuclear engineers.

The way in which employers see these instances is well illustrated by an exchange of views between myself and Arthur Bueche, vice president for research at General Electric Co., as described in the following quotation (*2*):

"Scientists who work in industry have an ethical responsibility to speak out on any research they are doing that could prove detrimental to the public, and scientific societies should take steps to protect the jobs of members who feel compelled 'to blow the whistle' on their employers," Dr. Alan C. Nixon, a former president of the American Chemical Society, says.

He spoke here this week at the annual meeting of the American Association for the Advancement of Science at an all-day meeting on ethics and the corporate scientist.

But Dr. Nixon's viewpoint was challenged by Dr. Arthur Bueche, vice president for research of the General Electric Company, who emphasized that the corporate scientist "owes loyalty to his employer" and cautioned that it is "often difficult to distinguish between those who are blowing the whistle and those who are just crying wolf." Dr. Bueche said that the primary ethical responsibilities of a corporate scientist or engineer were to perform "significant, relevant professional work," to protect trade secrets and to design and produce products that were "safe and effective."

If in the course of such work a scientist discovers what he believes are problems, Dr. Bueche added, he should first discuss the problem with associates and his management—and "be willing to resign" if he chooses to make a public attack.

This happened earlier this month when three General Electric engineers in California resigned from one of the company's nuclear divisions because they had reached the conclusion that nuclear energy represented a profound threat to mankind.

Their action focused renewed attention on a problem of increasing concern within the scientific community: how to separate what a scientist says as a concerned citizen from what he says as a scientist reporting scientific facts.

"We can tell them apart, but the rest of the public can't," Dr. Bueche said, adding that it can be "dangerous to the entire technical community" when a scientist "combines the role of activist and professional investigator."

But Dr. Nixon stated:

"I believe the scientist has responsibility to bring to the attention of the American public any problems in his field of science. Pollution is largely the result of chemistry, but our profession has not been very active in bringing this to the attention of the public. This came much later, usually the result of being caught by a government agency."

Dr. Nixon, a former industrial chemist, headed the American Chemical Society and, more recently, has been chairman of the Committee of Scientific Society Presidents. In that capacity, he has sought to encourage scientific societies to establish codes of ethics for corporate scientists, to investigate any cases of potentially harmful research brought to their attention by members.

Dr. Nixon said that the American Chemical Society had investigated 120 such cases in recent years* and "has expelled members who did something found to be unethical."

But he conceded that it remained "very difficult for a corporate scientist to speak out—often if he does, he is either fired or transferred to a less desirable assignment."

Since General Electric is a relatively responsible employer, this case demonstrates the huge gulf that exists between the employer and the employed as a result of the general employer tendency to consider the old English common law relation of master–servant as being the proper one for technical employees.

Existing Guidelines

The need for having some rules and regulations regarding employment somewhat better than common law has been evident for many years to concerned individuals in various professional societies. However, this concern was first translated into tangible form by the American Chemical Society (ACS) which has had a committee dealing with professional relations since the mid-40s. However, it was not until the uncertainties in the employment market of the late 60s and early 70s, which resulted in major multiple terminations of professionals, that the ACS was moved to formulate a set of minimum guidelines which were issued in mid-1971. They have since gone through three revisions, the most recent being endorsed in April 1975. Originally, the guidelines were directed exclusively to employers and were called "Guidelines for Employers" since it was felt that a previously adopted "Chemist's Creed" adequately covered the employee side of the equation. However, the most recent revisions have incorporated parallel guidelines for employees and are now called "Professional Employment Guidelines" (PEG).

Perhaps stimulated by the activity in the ACS, individuals in some of the engineering societies, particularly the AIChE (I. Leibson) and

* Most cases involve allegations of non-professional treatment by employers—very few involved "whistleblowers."

the N.S.P.E. (E. E. Slowter), began to formulate a similar document intended to cover both engineers and scientists. A joint committee was founded under the cochairmanship of Liebson and Slowter which produced a document entitled "Guidelines to Professional Employment for Engineers and Scientists" (GPEES), issued in January 1973.

Both sets of guidelines attempt to cover most of the types of problems that might arise in the employee–employer relationship and to set forth desirable conditions for employment. The scope of the ACS document can be comprehended by noting the headings in the PEG, which are: "Terms of Employment," "Employment Environment," "Professional Development," "Termination Conditions," and "Investigation of Unprofessional Conduct." Copies of PEG may be obtained from the ACS (*3*) and the GPEES from the EJC (*4*).

Problems with Guidelines

A major problem with the Guidelines, apart from somewhat naive expectations, is that the documents are not binding on either the employer or the employee. Acceptance of the Guidelines by an employer is completely voluntary. He does not have to announce that he is following them. Even if he does say he will follow them, he doesn't have to continue to do so in any particular case. Also, subscribing to the Guidelines may partially penalize the better employer who tries to abide by them because it puts him somewhat at an economic disadvantage viz-a-viz his less concerned competitor. The ACS Guidelines do have the virtue at least of being minimum criteria which the Society hopes employers of chemists and chemist employees will follow, and the ACS has set up a mechanism to measure performance by employers in cases of disputes or of multiple terminations. Individual chemists or chemical engineers can apply to the ACS under the Membership Assistance Program and have the ACS investigate a case of alleged unprofessional conduct on the part of the employer. If the allegation is deemed to have merit, attempts are made to reconcile the problem. Also, the ACS investigates all terminations which involve more than three chemists (or chemical engineers) and seeks to determine if the employer followed the Guidelines in connection with his treatment of his chemist employees. The cases are reported in *Chemical and Engineering News,* and the degree of compliance or noncompliance with the Guidelines is cited. The engineering societies, on the other hand, have set up no mechanism for measuring compliance. The guidelines are described as desirable goals, and although the engineers originally expected that they would be welcomed with open arms by employers and enthusiastically endorsed, this did not happen. As a result, their guidelines have had very little impact, and, as a matter of fact, the "ruling circles" of engineering societies have made it perfectly clear to employers that they are not going to go out of their way to bother them on the matter of guidelines. The following quotation describes their attitude (5).

The November (1974) *Astronautics and Aeronautics* of the AIAA, page 78, says " 'Teeth,' for Employment Guidelines Causing a Stir" in their member Newsletter. This has to do with the "Guidelines for Professional Employment for Engineers and Scientists" which has now been endorsed by 27 engineering and scientific societies. The AIAA Board gave qualified approval to the Guidelines —the qualification being that they remain "guidelines." The problem now is that of facing the necessity of enforcing the Guidelines. The Inter-Society Committee on Professional Employment Guidelines recognizes the necessity of treating them as minimum standards but it is up to the endorsing societies to provide a mechanism to see that the Guidelines are enforced. This is what is done by the ACS with respect to its own Guidelines. However, many of the engineering societies' boards of directors are dominated by employers' representatives who worry about their companies being placed under even this degree of constraint with respect to proper treatment of their employees. The AIAA Membership Committee "by consensus" held that the Guidelines should remain that and presumably no attempt should be made to see that they are enforced. Not surprisingly a subsequent engineer-industry conference demonstrated support for this point of view from industrial representatives.

Also in the same issue is an editorial by Al Cleveland, who is the AIAA's Director-Technical and VP-Engineering, Lockheed Aircraft Corp., in which he reports on "The Growing Dilemma of the Corporate Member" (AIAA has a practice of allowing corporations to be actual members of the organization, ACS has Corporation Associates). His dilemma seems to be brought about by the fact that the AIAA membership over recent years has been concerned about the employment conditions and mass unemployment of its membership and has been examining "such things as pensions and retirement plans, manpower planning, ethics, working conditions, salary standards, fringe benefits, and patent rights." He suggests that such preoccupation is bad because "it is . . . unreasonable to ask an individual to belong to and support a group, some portion of whose purposes and actions may be adverse to that individual's welfare; few politicians contribute to the opposition party." He apparently feels that these sorts of concerns are detrimental to industry; in fact, he says it is "putting it all in jeopardy." He suggests that AIAA should return to "technical matters" and leave "employment matters to be treated by other means." He implies that if this doesn't happen, corporations will pull out of the society and refuse to support any of its technical activities or allow employees to present papers, attend meetings, take part in the governance, etc.

I wonder if Mr. Cleveland is about to give up his U.S. citizenship because he doesn't approve of everything our government does?

Even when a society sets up a strong program for promoting compliance by employers with the voluntary guidelines for employment, it will have to expect that there will be a continuous attack on such a program by employers and by their representatives within the professional societies. This is happening now to the ACS. For instance, in 1975 a resolution was passed by the Committee on Professional Relations that *Chemical and Engineering News,* the organ of the American Chemical Society which goes to all members, should carry a story in the "Careers Issue" of the magazine each year listing companies with respect to how they had followed the guidelines in connection with multiple terminations or the treatment of individual members. It was also resolved that this story be repeated in three issues of the annual "Careers Issue." This request was denied by the editor of *C&EN,* and his action is now being supported by the Board of Directors. This sort of pressure against voluntary guidelines is to be expected.

It would appear that the only sensible course of action is to have the core provisions of the guidelines for employment enacted into law so that all professional employees and their employers will be under the same rules, and this sort of pressure on societies and their members will be removed. Although I have been advocating this for a number of years, one of the first general forums for presenting it occurred at a meeting at Alta, Utah in 1972. During the meeting, I headed a panel which presented a resolution endorsing legalized guidelines. These ideas subsequently found their way into a book by Nader, Petkis, and Blackwell called "Whistle Blowing" (*7*) and the book "Advise and Dissent" (*1*) by Primack and Von Hippel. They surfaced again at the AAAS meeting in Boston in February 1976 in the report of the AAAS on the social responsibilities of scientific societies. A listing of some of the items that such a law might include is given in the Appendix.

How will such a law come about? Obviously not by itself. Recently I wrote to Senator Edward Kennedy—a letter which is worth quoting in full—asking him to sponsor such a bill.

Dear Senator:

Thank you for giving me the opportunity to testify on the authorization for the National Science Foundation on the matter of Science for Citizens and for your gracious letter of March 4.

I would like to bring to your attention the matter which I did refer to in my testimony with respect to a law to improve the legal basis for the employee–employer relationship of professional employees. As I understand, and I am sure you as a lawyer are quite aware, the relationship is that of master–servant as defined in English common law. I think it would be very worthwhile to explore the prospects of changing this to provide legal basis for the employment relationship so that professional employees will be encouraged to exercise their responsibilities as informed citizens.

One might ask, "Why restrict it to professional employees and what is the legal basis for such restriction?" I think the answer to this is that:

• Professional employees generally are more apt to possess detailed knowledge and understanding of technical problems than do others.

- Generally employers consider the jobs that professionals do to be more sensitively related to the operations of the enterprise and, hence, are less willing to have such employees speak out in the area of their expertise than non-professional employees.
- As far as the legal basis for such identification, it seems to me that professional employees are defined as such in both the National Labor Relations Act with respect to their right to set up exclusive collective bargaining electorates and in the Wages and Hours Act where they are classified as exempt employees.

Not being a lawyer, I hesitate to undertake supplying the language for the framing of a bill, but in general I would think that the items that should be covered are:

- The necessity for there being set out clearly in writing the conditions of employment and legal obligations of the employee
- A standard non-discrimination clause
- The right of employees to work more than 40 hours per week if they wish but not regularly on demand (without compensation) if they do not
- Employees should have regular written reports on their adjudged level of job performance
- The special status of employees as professionals should be recognized
- Termination conditions should be spelled out

If you are interested in sponsoring a bill such as this, I would be very glad to assist in the drafting of the legislation or in any other way I can.

It will take the efforts of many people to accomplish the passage of such legislation. It will, of course, be opposed by most employers for the same reason that they refuse to endorse voluntary guidelines. It will cost them some money, and it will reduce their freedom of action. It is also opposed by labor unions who may see it perhaps as an infringement of their turf, which it is to some degree. However, even if all eligible professionals were unionized, I estimate as many as one-third to one-half of them would not be covered by union contracts because they would be classified as management employees under the rules of the National Labor Relations Board. The thrust for the enactment of guidelines will have to come from the people who are going to be affected directly—the professionals themselves—as Senator Kennedy pointed out in a reply to my letter. These matters have been presented to the various scientific and engineering societies through the Committee of Scientific Society Presidents and through the Association for Cooperation in Engineering and was briefly discussed at the joint meeting of these two groups in August 1976. The matter has also been discussed with Chris Stone of the University of Southern California who is the author of two forward-looking and innovative books designed to extend the law more effectively into areas that are not now adequately covered. He has expressed interest in contributing to the framing of an effective statement of a solution.

Recent law, which is only just now going into effect, should give a beneficial impulse toward the passage of legalized guidelines. I am referring to the so-called Science for Citizens Act which requires the National Science Foundation (NSF) to set up a program that will bring more scientists and engineers into the arena of public interest by providing mechanisms whereby they can lend their expertise to public interest groups and to influence public policy involving science and engineering matters. In testifying on the NSF appropriation bill before Senator Kennedy's Subcommittee of the Senate Committee on Labor and Public Welfare, I said the following (*8*):

However, it should be pointed out that there is a major problem in getting many qualified people to volunteer for this effort. As you know most engineers work for industrial organizations while a substantial and rising proportion of scientists do likewise. Industrial managements are extremely reluctant to have technical employees speak out on technical matters except through the management structure. This was emphasized at the recent meeting of the American Association for the Advancement of Science in Boston on a panel on "Ethics and Corporate Scientists." I expressed the point of view that scientists who work in industry have an ethical responsibility to reveal information that they have which they feel should be divulged for the protection of the public. This view was challenged by Dr. Arthur Bueche, vice-president for research at General Electric Company, who emphasized that a corporate scientist primarily "owes loyalty to his employer." Dr. Bueche felt that if a scientist or engineer wished to speak out, his proper course was to discuss it with his management and, failing satisfaction, resign if he felt he had to go further. I do not think that Dr. Bueche is right; it should not be necessary for an industrial scientist or engineer to resign in order to contribute to public policy or safety. But a mechanism must be created in order to protect such individuals.

It was probably not appreciated by the drafters of the bill that it would be very difficult to get scientists and engineers employed in industry to come forward with information on public policy concerns in the absence of some protection for their livelihoods.

The passage of a law legalizing guidelines, of course, will not remove the necessity for professional societies to be active in the field of protecting their members. No law is self-enforcing. It is to be expected that unscrupulous employers will seek to evade the provisions of the law. Professional societies should have mechanisms available to their membership at all times which will apprise them of their rights under the law and be able to give them advice on particular issues, and should be willing to provide legal aid funds in case it is necessary for them to insist on compliance with the law. Also, the professional societies can help their members in those cases when the law, as all laws do, breaks down and the member is forced to seek other employment as a result of psychological pressure or subtle acts of discrimination. However, given the law and the assistance of the professional societies, professional employment in the United States can be expected to be more productive, more stimulating, and more rewarding.

Literature Cited

1. Primack and Von Hippel, "Advice and Dissent," Basic Books, New York, 1974.
2. Wilford, John Noble, "Scientists Discuss Dual Loyalty on Job," *N.Y. Times*, Feb. 22, 1976.
3. "Professional Employment Guidelines," June 1976, American Chemical Society, 1155 16th St., NW, Wash., D.C. 20036.
4. "Guidelines to Professional Employment for Engineers and Scientists," Jan. 1973, Engineers Joint Council, 345 E. 47th St., N.Y., N.Y. 10017.
5. *The Vortex*, Calif. Section of the ACS, Berkeley, Calif., March 1975, p. 52.
6. "Scientists in the Public Interest: The Role of Professional Societies," The American Academy of Arts and Scientists, Boston, 1972.
7. Nader, Petkis, Blackwell, "Whistleblowers," Grossman Publishers, New York, 1972.
8. *NSF Senate Authorization Hearings*, Special Subcommittee on the NSF of the Committee on Labor and Public Welfare, March 3, 1976.

RECEIVED December 16, 1976.

Appendix

Outline of Proposed Legal Guidelines for Employers of Professional Employees

In connection with the employment of such employees, the following conditions will prevail:

I. Conditions of employment shall be fully described in writing to a prospective employee.

II. Legal obligations of the employee to the employer must be clearly set forth in an employment agreement.

III. Employment shall be based solely on competence and ability to perform assigned responsibilities without regard to factors of age, race, religion, political affiliation, or sex.

IV. Such employees may not be regularly scheduled to work more than 40 hours per week unless recompensed at the rate of 1.5 times the normal salary for hours in excess of the 40 hours.

V. Such employees must be provided with opportunities and facilities for working more than 40 hours per week if they so desire.

VI. An employer must keep such employees informed of their judged level of job performance by means of confidential written annual records which must be attested to by the employee with copies provided.

VII. The Judgment of a professional employee's performance should involve input from his professional peers by means of confidential questionnaires.

VIII. Employers should allow such employees opportunities to maintain professional expertise through attendance at professional meetings equivalent to at least one week per year and courses of study and leaves of absence for professional study equivalent to two weeks per year.

IX. Employee–inventors shall participate to the extent of at least 10% in the income generated by their inventions. Patents obtained by an employee resulting from inventions developed on his own time, outside his assigned field of work, will be the property of the employee.

X. Employees practicing in professional fields where advancement is dependent upon publication will not be inhibited from so doing except through contractual arrangements.

XI. The right of professional employees to participate in the activities of their professional societies will be recognized.

XII. Employees will receive notice of intended termination one month plus two weeks for each year of service before such termination will take place, except that severance pay may be offered in lieu of notice beyond two weeks.

XIII. Terminated employees will be provided reasonable assistance in finding another position.

XIV. Employees terminated due to budget cuts or a reduction in force will have first priority in rehiring for two years beyond their termination date. Employees terminated with a minimum of 10 years of service shall have fully vested pension rights.

XV. Employees with a minimum of 10 years of service may not be terminated except for fully documented cause confirmed by two levels of management above his immediate supervisor, confirmable by employees' peers.

XVI. Health and insurance plans shall be continued for a terminated employee for a minimum of one month beyond termination plus one week for each year of service.

XVII. Upon termination, proportional vacation rights shall be exercised by the employee, either in time or money.

XVIII. Upon termination, the employee shall be given custody of his central personnel file. Records remaining with the employer will consist simply of the name and address of the terminated employee and the period of service. Upon termination, an employee shall be delivered a document by the employer defining his residual rights in inventions, patents pending, and possible publications.

Discussion

Q. What can local sections do with respect to employers of chemists in their territory?

A. I think that what they should do is to write to each employer, send him a copy of the Guidelines, and bring to his attention the fact that they do exist and follow up when they get an answer.

Q. The problem is how far you can go because I received a memorandum from the ACS legal counsel stating that we would be very limited in how far we could go.

A. That was a very garbled communication from the Committee on Professional Relations. It was a most ill-advised thing to do, and I wrote and told them so. If they want to communicate with the local sections, they should say in clear and unambiguous terms what they advise them to do. That was just a cloud of double talk that was sent out, and I am not surprised you reacted the way you did because I thought that was what would happen. I think that you should write to your employers and say that the Guidelines exist. Local section Professional Relations Committees should not get too involved in interacting with any situation but should concentrate on getting information back to Washington and let it be handled from there, because you could get out of your depth very rapidly. The information gathering function is most important, and that is the thing that has to be done in these cases.

Q. Shouldn't the local section Professional Relations Committees visit employers to discuss the Guidelines?

A. I think that it would be much more advisable to write and say that you have just received copies of the latest edition of the Guidelines and that as a local section you are very much in favor of them rather than commenting by word of mouth. I think that it is much better to put the local section on record in writing with respect to their position and then to follow up by visits if it seems desirable.

Q. How is professionalism defined?

A. Professionalism is defined by law in the National Labor Relations Act.

Q. But what is your definition of a professional?

A. I think that a professional is somebody who has had a course of training such as that offered by a recognized university or college in a discipline or group of disciplines and that he or she practices in an area that requires that sort of background in order to hold the job.

Q. Are teachers professionals?

A. Yes.

Q. Will Congress legalize guidelines when there is an existing alternative way of achieving the same goal?

A. Well, the question relates to whether Congress will pass a law unless you can show clearly that there is no (or little) possibility of doing it otherwise. I suppose you are referring to the possibility of professionals' forming a union and then getting bargaining rights under the NLRB and writing the guidelines into a contract. That of course is a possibility; it can be done and may be a very good way to go. On the other hand, a large fraction of all professional employees (probably one-third of them) are not eligible to be in bargaining groups. They would be classified as management employees and therefore not eligible for inclusion under a collective bargaining contract, so that these people would not be covered. Such people often feel as though they are meat in a sandwich—that the management above has its own private arrangements and a union below has its own private contractual arrangements, and they are in the middle with no collective protection and no possibility of getting any unless the NLRB is amended to include management. Also, many chemists work for small companies where they are employed as chemists, but the management considers them as part of management. Although they are not strictly management, the chances are that they would not be included in a bargaining unit. Many others do not want to join unions. So there are a lot of people who won't be covered unless Congress passes a law that protects all professionals. Such a law would enable us, in a sense require us, to come forward as good citizens to help protect the public from chemical insults and to aid in the development of public policy involving science and technology.

Q. Does the ACS publish the names of employers that follow the Guidelines?

A. What mostly has been published are violations of the Guidelines in multiple terminations (layoffs). Of course, if a company follows the Guidelines in connection with a multiple termination, that is reported. However, there was an article in *C&EN* a couple of years ago about companies that had avoided layoffs and some complimentary things were said about them. Unfortunately, soon after that, a couple of them fell off their pedestal.

7

Rights of Chemists, Employers, and Professional Societies in Layoffs and Other Serious Grievances

WILLIAM J. BUTLER

Hanson, O'Brien, Birney, and Butler, 888 17th St., N.W.,
Washington, D. C. 20006

The rights of chemists, employers, and professional societies are discussed in relation to layoffs and other grievances where the Labor Management Relations Act, the Fair Labor Standards Act, the Age Discrimination in Employment Act, the Civil Rights Act of 1954, and the Occupational Safety and Health Act of 1970 apply. Instances of political discrimination, blacklisting, employee entrapment and dismissal, and no-switching agreements are also included in the discussion.

In the area of the rights of chemists in layoffs and other serious grievances, there is no question that the employee is getting a chance to say more, owing to the assistance of legislation such as Title VII of the Civil Rights Act of 1964 and the courts, which are now starting to play a role. In these economic hard times the threat of a layoff is very real, both to the factory line workers and the white collar or professional employees.

It seems as though in times of recession, research and development is usually one of the first areas where belt tightening occurs. Of course a more basic reason why layoffs occur, whether they be economically justified or not, is that fewer and fewer American workers are self-employed. As a result of a steady increase over the years, almost 90% of the labor force are wage or salary earners. At the same time, membership in labor unions has remained fairly stable since World War II, with the proportion of organized employees in the labor force actually declining. This is significant since membership in a labor union may afford various protections to a worker by virtue of certain labor laws and because unions are organized and can negotiate employment contracts from a position of relative strength. Finally, as the power of corporate employers has increased, the bargaining power of non-union employees has decreased in some areas.

Three factors are at work which make the issues of layoffs important to every professional wage earner. First, layoffs are obviously more prevalent in times of economic recession and undercapacity. In fact, since the last American Chemical Society (ACS) meeting in 1975, 29 separate layoffs have been brought to the attention of the ACS Employment Aids Office. Each layoff involved from three to as many as 31 chemists or chemical engineers. (A layoff for our consideration requires only that three people be laid off rather than the masses of people required by the older definitions). Distressing as these figures are, they are substantially below those for the years 1969–1971.

Second, most professionals do not enjoy either the strength in numbers which a labor union offers or the full protection of such labor legislation as the National Labor Relations Act or the Fair Labor Standards Act. NLRA prohibits unfair labor practices on the management's part aimed at organized labor or aimed at unorganized workers "engaged in concerted activity for their mutual aid or protection"—a concept that has come to be construed expansively by the National Labor Relations Board. The Fair Labor Standards Act sets minimum standards for labor conditions, such as minimum wage and overtime provisions, but it exempts from its coverage bona fide executives, administrators, and professional employees, the categories into which most chemists fall.

The only known increase in unionization among professionals to date has occurred in the context of governmental and institutional employees and, particularly, in the field of education. There, the National Education Association, the American Federation of Teachers, and the American Association of University Professors are all competing for the right to represent the professional academic employees on American campuses. In the context of the private business firm, however, either the number of professional employees is too small or the economic interests are too diverse to give them any real economic influence vis-a-vis their employer in a bargaining situation.

The third factor, which should be of some concern to the professional chemist in terms of his vulnerability to a layoff, is the near absolute lack of legal protection (other than what may be contained in the employment contract itself) against termination, be it for a just cause or not. The classic statement of an employer's right in this area is found in the 1884 Tennessee Supreme Court opinion in the case of Payne vs. Western & Atlantic Railroad (*1*) where it was said "all employers may dismiss their employees at will for good cause, for no cause, or even for cause morally wrong without thereby being guilty of legal wrong."

The story of how this principle became the accepted legal doctrine in this country is a curious one. Professional employees traditionally had been employed under oral or, in some cases, written employment contracts that ran for an indefinite term. Under the English common law, when no particular term of employment was stated, the hiring was presumed to be for one year's service. If the employment continued beyond one year, it was thereafter terminable only at the end of an additional year. Other than this power unilaterally to terminate at one-year intervals, both parties, under the English rule, would be bound as long as they both remain satisfied with the performance and the working conditions. In other words, except for the power to terminate automatically on the anniversary of the date on which the contract was made, both parties could terminate only for a good cause shown.

In the United States the rule was quite different. Unless the life of the employment contract is specifically spelled out, such a hiring is deemed terminable at will by either party. Whether the contract provides for payment at stated intervals, such as every two weeks, once a month, etc., or even characterizes the employment as permanent, has no effect upon the legal result. The contract is terminable at the will of either party.

An interesting fact in regard to this American rule is that it is the result of a single scholar's interpretation—that of H. G. Wood in his 1877 treatise, "Master and Servant." After publication of this treatise the courts simply began to apply Wood's interpretation, at first merely citing Wood as an authority, then later citing intervening cases which had cited Wood. Thus, we have the law as it stands today. This doctrine —that employment contracts for an unspecified term are terminable at will—was adopted without any penetrating legal analysis. Perhaps the real rationale for this doctrine was that it fit the economic and social context at a time when management was largely unregulated and the economy was guided by the unseen hand of laissez-faire and *caveat emptor*.

It is doubtful, however, that this rule is justified today when full employment and aid to the unemployed are avowed objectives of social policy, and job security has almost become a fundamental right in labor law, exemplified in 82% of modern collective bargaining agreements where "for cause" and "just cause" restrictions upon the employer's right to terminate an employee are found. Of course, merely because the employer's power to terminate at will—an absent employment contract provision to the contrary—is the general rule of law does not mean that it has been imposed in all situations where employees are discharged. A few courts have suggested that an employee's promise to perform services for an unspecified period is legally sufficient either to bind the employer to an implied promise to discharge only for cause or to create the option on the employee's part to remain as long as the work shall be satisfactory and needed.

Other courts have reinstated discharged employees under a theory somewhat similar to unconscionability. The Doctrine of Unconscionability and its use in preventing bad faith on the part of one party to a contract, however, has been restricted largely to non-employment cases involving sales, wherein a consumer, usually uneducated, is unwittingly duped by a conniving sales representative.

There are other cases where courts have found unemployment contracts to be void, such as an adhesion contract. In an adhesion contract one of the parties has no real choice. He must take it or leave it, despite the fact that the contract contains outrageously onerous provisions. Unfortunately, the courts generally use this doctrine to invalidate existing contract terms and not to create new, implied obligations. Thus, although courts have occasionally invalidated existing employment contract provisions, and even less frequently have held that job security implicitly exists in open-ended employment contracts, very few remedies, until recently, have been available to an employee discharged without notice, without cause, without severance pay, and without rehiring privileges.

The vulnerability of the professional employee to unfair termination has finally been recognized, however, and is illustrated by abusive discharge cases, of which I shall now mention three. In a 1959 California case called Peterman vs. Teamster Local 296 (*2*), the court reinstated an employee who had been fired for failing to commit perjury when solicited to do so by the employer. In that case the court said:

> it would be obnoxious to the interests of the state and contrary to public policy and sound morality to allow an employer to discharge an employee, whether the employment be for a designated or unspecified duration, on the ground that the employee declined to commit perjury, an act specially enjoined by statute. The threat of criminal prosecution would, in many cases, be sufficient deterrent upon both the employer and the employee, the former for soliciting and the latter for committing perjury. However, in order to more fully effectuate the state's declared policy against perjury, the civil law, too, must deny the employer his generally unlimited right to discharge the employee whose employment is for an unspecified duration when the reason for this dismissal is the employee's refusal to commit perjury (*3*).

Although this case involves extreme hardship, it allows us to get new laws of this type on the books.

In the 1973 case of Frampton vs. Central Indiana Gas Co. (*4*), the court reinstated an employee who had been fired for filing a workmen's compensation claim. The court said:

retaliatory discharge for filing a workmen's compensation claim is a wrongful, unconscionable act and should be actionable in a court of law. Although we know of no other cases in this or any other jurisdiction holding that such a discharge is actionable, there has been a parallel development in landlord and tenant law. Courts in several jurisdictions have held that retaliatory evictions offend public policy.

The court went on to say that the retaliatory discharge and retaliatory evictions are clearly analagous and then said,

we agree with the Court of Appeals that under ordinary circumstances, an employee, at will, may be discharged without cause. However, when an employee is discharged solely for exercising a statutorily conferred right, an exception to the general rule must be recognized (*5*).

Finally, in the 1974 New Hampshire case of Monge vs. B. B. Rubber Co. (*6*), the court in its opinion said:

plaintiff claims that she was harrassed by her foreman because she refused to go out with him and that his hostility, condoned if not shared by defendant's personnel manager, ultimately resulted in her being fired.

The court here stated:

the law governing the relations between employer and employee has similarly evolved over the years to reflect changing legal, social and economic conditions. In this area, we are in the midst of a period in which the pot boils the hardest and the process of change the fastest. Although many of these changes have resulted from the activity and influence of labor unions, the courts cannot ignore the new climate prevailing generally in the relationship of employer to an employee. . . . in all employment contracts, whether at will or for a definite term, the employer's interest in running his business as he sees fit must be balanced against the interest of the employee in maintaining his employment, and the public's interest in maintaining a proper balance between the two. We hold that a termination by the employer of a contract of employment at will, which is motivated by bad faith or malice or based on retaliation, is not in the best interest of the economic system or the public good and constitutes a breach of the employment contract (*7*).

These cases represent a new trend in the law by making inroads into what has been the management's complete right to discharge an employee without good cause. However, in each of these cases, two elements were present. (1) All three involved non-work related discharges, for the threat of being fired was first used by the employer to extort or coerce the employee into committing an act outside the parameters of his legitimate job description. (2) In each case, a strong public policy weighed in favor of the employee's actions—namely, in Monge, an employee's right to associate or not to associate with persons of her choice; in Frampton, workmen's compensation; and in Peterman, criminal penalties for perjury. These cases, therefore, may be more of a vindication of society's interest than a vindication of an employee's right to implied job security.

This notion of implied job security, accruing to an employee from a source outside the stated terms of his employment agreement, has been given its most authoritative endorsement by the U.S. Supreme Court in the 1972 case, Perry vs. Sindermann (8). In that case Sindermann had taught in the Texas college system for 10 years, having taught for four years at Odessa Jr. College immediately prior to his dismissal without cause. Odessa College had no tenure system, and the plaintiff had no formal, contractual right to job security. The school's faculty guide stated, however, that a faculty member's job was safe as long as a teacher's services were satisfactory and he was cooperative. In 1969 Sindermann was fired with no official explanation, no hearing, and no right to appeal. He then sued in federal court for reinstatement, arguing that the decision not to rehire him was based on his outspoken criticism of the college administration, thus infringing on his right to free speech, and that the College Board of Regents' failure to grant him a hearing violated his 14th Amendment right to procedural due process. The Court agreed, at least with the due process claim, and although the Court did not rule that Sindermann had a right to continued employment, it did remand the case to the trial court and ordered that he should have the opportunity to prove an implied contractual right to employment.

In what may be a landmark effort of courts to strike a new balance between employee and employer rights, the Supreme Court said:

> A written contract with an explicit tenure provision clearly is evidence of a formal understanding that supports a teacher's claim of entitlement to continued employment unless sufficient "cause" is shown. Yet *absence of such an explicit contractual provision may not always foreclose the possibility that a teacher has a "property" interest in re-employment.* For example, the law of contracts in most, if not all, jurisdictions long has employed a process by which agreements, though not formalized in writing, may be "implied." Explicit contractual provisions may be supplemented by other agreements implied from "the employer's words and conduct in light of the surrounding circumstances" (9).

Thus, for the first time, an employee's right to job security does not stand or fall on the basis of the formal terms of the employment contract. Although courts will not recognize and protect one's expectancy of continued employment or expectancy of discharge only for cause, an employee may have a right to continued employment or a right to require a showing of cause to justify his dismissal if such rights are recognized in the policies, informal statements, or practices of the employer.

A word of caution is in order. This case is easily distinguishable from that involving layoffs of privately employed chemists and chemical engineers. Sindermann was a public employee, and his claims alleged a denial of his constitutional right by a state agency. Furthermore, the Supreme Court held that even if Sindermann were able to prove an entitlement to job security, he would not necessarily be reinstated. Instead, the college would merely be required to show good cause for his dismissal, and courts historically have been fairly acquiescent in recognizing good cause for dismissal in a private employer's claims of economic necessity. Thus, although the long-shut door to professional employees' rights in job security has now been opened, most courts still will not overturn a private employer's decision to discharge an employee except under certain conditions:

• A court will clearly order reinstatement if an employee's dismissal violates the terms of his employment contract.

• A court will overturn an employee's dismissal where it violates the 1964 Civil Rights Act, which makes it unlawful for an employer to "discharge any individual . . . because of such individual's race, color, religion, sex or national origin." (A chemist who is employed by a government contractor receives additional protection by virtue of Executive Order 11246, which provides that government contractors found to have engaged in employment discrimination may have their contracts terminated.)

• A court will overturn the dismissal of a chemist between 40 and 65 years of age where it violates the Age Discrimination in Employment Act of 1967, which makes it unlawful "to fail or refuse to hire or to discharge any individual with respect to his compensation, terms, conditions, or privileges of employment because of such individual's age."

• A court might overturn the dismissal of an employee where that employee, in addition to incurring obligations or surrendering certain rights in exchange for employment, surrendered additional rights or incurred additional obligations in exchange for permanent employment. An example of this would be an employer's receiving extraordinary benefits from hiring a particular employee—e.g., when an employee agrees to surrender certain tort claims against the employer. It was common early in this century for the employer to extract from the employee, as a condition of employment, a waiver of his common law right to recover damages for injuries resulting from industrial accidents. Another example is where the employer solicits and the employee makes special contributions to the business.

• Courts will also look for any special reliance by the employee; for example, where an employee changes jobs at some personal sacrifice in order to work for his new employer, or where a businessman sells his business contingent upon being hired by the new owner.

• Another factor which may sway a court, particularly since the Sindermann case, is whether the "common law of the job" indicates a right of employment security. In other words, is there a handbook of work rules, an implied promise, or an oral statement or memo to the worker, or anything else supplemental to the contract which creates an enforceable right of job security for the employee?

• Finally, the court will look to the question of the employee's longevity of service with that particular firm. This does not mean that those workers who have served longer on the job have greater rights than those who haven't. It does mean, however, that, particularly where a dismissal would work an extreme hardship because of an employee's loss of deferred retirement compensation, or where a firm temporarily lays off an employee to prevent him from receiving accelerated increases in benefits which come with uninterrupted longevity, a court will consider longevity in favor of the discharged employee.

This, then, is the changing state of the law regarding dismissal of professional employees who have not contractually bound their employer to retain them for a given term. Reform is visible on three fronts. First, courts are paying closer attention to the contract–law concept of implied promises—i.e., a promise of job security not found in the employment contract but which arises from circumstances of employment, the company's "common law," etc.

Secondly, courts are more responsive to claims that a particular discharge may be sued upon in court, not because it violates an enforceable contract right of an employee, but because it shows bad faith and undermines an overriding public policy. The Petermann, Frampton, and Monge cases, as well as cases founded upon any of the employment discrimination acts, illustrate this. There is also a growing species of tort known as the *prima facie* tort, where offending parties may be called to answer for acts which are not wrongful in themselves but are wrongful when done out of spite or malice. Examples include the erection of "spite" fences, drilling a well on your property solely to cut off another's underground water supply, and "abuse of process," where a plaintiff uses lawful procedures of the court and sues another merely to harass him.

Thirdly, there is much agitation for the creation by state legislatures of a private right of action for the wrongfully discharged professional employee. Under California law today, fair notice is required of an employer prior to an employee's termination, and an employer is further prohibited from using such tactics as demotion, failure to promote, poor work assignments, frequent and undesirable transfers, and general harrassment. Missouri has a law requiring a statement of the reasons for an employee's discharge, upon request.

Although the future for the rights of professional employees in this area is rather bright, for the present one must negotiate and demand that provisions for employment security be specifically incorporated into the terms of the employment contract. These cases where the court looks beyond one's rights under the contract in granting relief to a terminated employee remain exceptions, not the rule.

A more prophylactic protection against the economic dislocation of a layoff is the pension and retirement fund. The exact nature of the rights and benefits offered by a private pension plan, of course, depends upon the particular provisions of that plan, and great variety exists. Some states regard a retirement plan as wages withheld in order to induce continued faithful service on the employee's part; therefore, the state interprets the plan as delayed compensation for services rendered. Under this interpretation, upon dismissal an employee may enforce his claim to whatever amount has been accumulated for him in the retirement fund. Other states take the view, particularly where the plan does not call for employee contributions, that retirement benefits are gratuities to which employees do *not* have an enforceable entitlement. Your attorney should explore the law of your state on this question before assuming that some amount your employer has set aside as retirement benefits can actually be taken with you upon dismissal.

Perhaps more important than state law, however, in determining one's legal rights under a pension plan is ERISA—the 1974 Employee Retirement Income Security Act. ERISA, by altering the requirements which employee retirement benefit plans must meet in order for an employer's contributions to qualify for tax deductibility, has standardized private pension plans in ways most favorable to covered employees. Tax savings and protections are also available under either a "qualified" group retirement plan or an H.R. 10, "Keogh," or I.R.A. plan wherein an individual may set up a tax-deferred retirement fund for himself.

Summary

It behooves every professional chemist to recognize, in terms of job security, that he is less well protected than workers covered by approximately 80% of the collective bargaining agreements in this country. At least under those contracts, discharge of an employee must rest upon a showing of a specially enumerated good cause. It further behooves professional chemists to realize that nine times out of 10, job security only comes from the terms of your employment contract. One's enlightened self-interest, therefore, demands tougher contract negotiation to upgrade protection against layoff. In conjunction with that effort, an employee must explore for himself or through an attorney, the law in his state regarding abusive discharges, vesting of retirement benefits, unemployment compensation, and rehiring privileges. The job security of the professional employee may be largely hostage to the whim of the employer. There is no other situation where the maxim, "the law helps him who helps himself," is more appropriate.

Literature Cited

1. Payne vs. Western and Atlantic Railroad, 81 Tenn. 507,520 (1884); overruled on other grounds, *Hutton* vs. *Watters,* 132 Tenn. 527 (1915).
2. Peterman vs. Teamster Local 239, 344 P.2d 25 (1959).
3. Ibid., 27.
4. Frampton vs. Central Indiana Gas Co., 297 N.E.2d 425 (1973).

5. Ibid., 428.
6. Monge vs. B. B. Rubber Co., 316 A.2d 549 (1974).
7. Ibid., 550–551.
8. Perry vs. Sinderman, 408 U.S. 593 (1972).
9. Ibid., 601–602.

RECEIVED August 26, 1976.

8

A Union's Effect on the Legal Obligations of Chemists and Employers

DENNIS CHAMOT

Council of AFL–CIO Unions for Professional Employees,
815 16th St., N.W., Washington, D.C. 20006

The average chemist works at the sufferance of his employer. He can be fired for any reason not covered by anti-discrimination statues, cannot bargain over salary or fringe benefits, and cannot openly protest undesired assignments without fear of losing job or promotional opportunities. With the formation of a union, the employer loses a good deal of flexibility. The law requires him to bargain in good faith with the union, being bound by a collectively bargained contract. The chemist is also bound by the contract negotiated by his elected representatives. Formal means are available for settling differences based on National Labor Relations Board rulings and past court decisions.

The typical professional today, and chemists are no exception, differs in many ways from the professionals of the past. The true professional was an independent, self-employed individual who offered his services to clients. He sold his expertise as a commodity that was valuable and in demand. For these reasons and because fully trained professionals such as physicians and attorneys were relatively rare, these people possessed a great deal of individual bargaining power. The legal obligations of the professional and his client were derived from the contract, written or oral, that was signed by the two parties. The professional agreed to perform the job satisfactorily according to the terms of the contract, and in return, the client was required to pay the fee agreed upon and to provide whatever other material support required. The situation today is vastly different for most professionals primarily because most professionals today are employees. They do not accept clients; they are hired by employers. They have lost their independence and with it much of their ability to protect themselves.

Few chemists work under detailed employment contracts. As such, they have many obligations and few protections. What contractual agreements do exist usually deal with peripheral matters and are heavily in the employer's favor. For example, the chemist frequently must agree to assign all inventions to his employer as a condition of employment. The chemist is subject to many employer-dictated rules and regulations, not because there is any legal basis, but because he is an employee by sufferance, and as such, he has little job security. His lack of individual bargaining power leaves him with little or no input in formulating those policies and rules.

Furthermore, the employee can be fired at any time, for any reason, except for those situations covered by law—discrimination based on race, sex, age, religion, or union activity. Indeed, since most chemists are exempt employees under the Fair Labor Standards Act, the employer does not even have to pay them for overtime work.

In a curious way this dual system of saddling the employee with stronger constraints than the employer is reflected in the ACS's own Professional Employment Guidelines. For example, under Section 1, "Terms of Employment," the chemist is "obligated to honor an offer of employment once accepted unless formally released after giving adequate notice of intent." The equivalent section for the employer states: "The employer is obligated to honor a written and accepted offer of a position. *If unable to honor it,* the employer *should* provide the chemist with equitable compensation." In other words, the chemist is given no alternative, but the employer has a great deal of discretion.

The foregoing deals with the relations between an individual chemist and his employer. The problems exist for the most part because of the vast difference in bargaining power possessed by the two parties. An employee union changes the situation markedly, in part because of legal restrictions placed upon the employer both by the collective bargaining agreement and by applicable labor laws.

It is useful at this point to define terms. The Labor Management Relations Act, commonly known as the Taft-Hartley Act, doesn't use the word "union" but does define the term "labor organization" as "any organization of any kind, or any agency, or employee representation committee or plan, in which employees participate and which exists for the purpose, in whole or in part, of dealing with employers concerning grievances, labor disputes, wages, rates of pay, hours of employment, or conditions of work." This is a broad definition. Essentially any voluntary group of employees acting together to deal with their employer over wages, working hours, or working conditions is, in fact, a union.

Section 7 of the Taft-Harley Act recognizes the right of employees to join together for the purposes of bargaining collectively with their employer. Section 8 further provides, in part, that it is an unfair labor practice for the employer to interfere with the employee's right to form or to work for a labor organization. In other words, a chemist cannot be fired for trying to form or for joining a union; he cannot be discriminated against for carrying on union activities; and the protection begins when the activity begins, not after the union is certified.

The term, "unfair labor practice," is used in the Taft-Hartley Act. Violation of any of the specific practices stated in the Act are subject to remedial action ordered by the National Labor Relations Board. The Board has the authority to go to court to ensure that its decisions are carried out. If an employee who is fired or otherwise discriminated against so charges, the burden of proof is on the employer to show that such action is completely unrelated to union activity.

One other item in the Taft-Harley Act is particularly pertinent—the obligation placed on both the employer and the union to bargain in good faith. The employer is not obligated to agree with every suggestion and demand of the union, but neither can he dismiss them. The result of the process is a written employment contract that is truly bilateral and not employer dictated. This system closely resembles the system of the past where professionals possessed significant individual bargaining power, though here a group is involved. The signed written contract is binding on both parties. Hence, it may be of value to outline some of the areas usually covered and also to look at what happens when a disagreement on interpretation or an outright violation of the contract occurs.

The statutory scope of bargaining—"wages, hours, and other conditions of employment"—is very broad. Bargaining over items contributing to employee income includes rates of pay and kinds of pay increases to be granted (straight percentage; percentage based on consumer price index changes; merit pool only, with allocation determined either solely by management or with peer input; across the board dollar increases, or some combination of these) as well as bonus plans, stock purchase plans, and extra compensation for inventions and valuable ideas. Also subject to bargaining are vacations, holidays, employer contributions to health and life insurance plans, and pension plans. The union participates in setting these policies previously decided solely by management.

An important subject is layoffs. The union and management may bargain over criteria for dismissal, amount of notice, amount of severance pay, whether or not benefits like insurance policies will remain in effect, and important recall provisions. The customary treatment of chemists has been to ignore recall possibilities. Layoff is equivalent to permanent separation, but this need not be. If the layoff is caused by economic considerations (regardless of the excuse given) and is not a result of poor performance, the chemist should be returned to work when the economic situation improves. This can be covered by the collective bargaining agreement.

A great many other subjects may be brought up during contract negotiations, but I would like to discuss just one more. Any contract, no matter how skillfully drafted, is subject to various interpretations. Some may be deliberate attempts to wring more out of the contract than was put in. More likely, there may be honest differences of opinion. A mechanism must be available for resolving problems in either case.

Most collective bargaining agreements will include a formal grievance procedure. A grievance is nothing more than a complaint that some part of the contract has been violated. Since the contract covers broad areas of compensation, fringe benefits, and working conditions, the availability of a formal, binding complaint procedure that is not controlled by management gives the employee protection that he otherwise would not have. Indeed, any good grievance procedure will include third-party arbitration as the last step, which places the final decision in the hands of a neutral outsider.

Many checks and balances are built into this system. The goal is compromise and the prevention of absolute dominance by either side. Government entities stand ready to enforce the law no matter which side brings the complaint. The first is the National Labor Relations Board, mentioned earlier in the context of protecting employees' rights regarding union activity. The Board is charged with administering the National Labor Relations Act. Although the Act sets general policies and spells out specific unfair labor practices for both employers and unions, it cannot cover all possible grievances that may arise. Interpretation and enforcement are left to the Board. After 40 years of handling cases and rendering decisions, the NLRB has built up a large body of precedents dealing with unit determination, refusal to bargain, discrimination for union activity, access to information needed for bargaining, and so on. It has the authority to issue binding orders and can secure enforcement through the federal courts. On the whole, the system works remarkably well. Unfortunately, it is not available to all. The NLRB does not have jurisdiction over public employees—e.g., chemists who are employed by state universities or federal or state agencies. Many of these, however, are covered by some kind of collective bargaining law which permits organization and at least limited bargaining.

Federal employees, for example, are covered by executive orders going back to 1962 which require federal agencies to recognize employee organizations. However, these unions cannot bargain over pay scales which are set by Congress. State and local employees may be covered by individual state laws which vary from state to state. Complaints can be brought to state employee relations commissions.

Public sector bargaining is fairly recent and is not yet universal. The situation is still developing. Where bargaining is permitted, even if the scope of bargaining is statutorily limited, the role of unions is the same as in the private sector—to increase employee bargaining power.

Whether in the public or private sector, employee–management relations are governed by rules. In the absence of a union, the rules are unilaterally set by management. Legal restraints on the employer are relatively minor. With a union the rules are jointly negotiated between employer and employees and take the form of a written, enforceable contract. The locus of rule-making authority is shifted; simultaneously, the possibilities for appeal to outside agencies are increased.

The applicability of this argument in favor of unions for professional employees is especially important to chemists. The number of chemists in unions now is relatively small, but the number of other professionals —actors, musicians, nurses, doctors, journalists, school teachers, college professors, and engineers—is large, approximately three million (*1, 2*). Professional chemists can be expected to fall into this category by following the trend and simply because unions do indeed change the legal relationship between the employer and the employee, distinctly in favor of the employee (*3*).

Literature Cited

1. Chamot, D., "Professional Employees Turn to Unions," *Harv. Business Rev.* (May-June 1976).
2. U.S. Department of Labor, "Directory of National Unions and Employee Associations," Bureau of Labor Statistics, Washington, D.C.
3. Chamot, D., "Scientists and Unions: The New Reality," *Amer. Federationist* (Sept. 1974).

RECEIVED August 13, 1976.

9

Societal Responsibility of the Practicing Chemist

ALBERT J. FRITSCH

Center for Science in the Public Interest, 1757 S Street, N.W., Washington, D.C. 20009

Chemists have helped fashion modern society. Their scientific procedures and acquired knowledge have societal content, especially in the area of toxic chemical substances. This content extends to choices of research topics, information flow, and public policy making. Quite often the chemist is able to alert the public to possible dangers in the use of certain chemicals. If whistleblowing is the only effective recourse, proper procedures should be followed. As a member of a professional group, the chemist is becoming conscious of the need to champion good hiring practices, working conditions, job security, and pension policies. This awareness is broadening to procedures for securing grants, gathering data, chemical applications, and the method for revealing published results. Examples of concrete problems are included.

Chemists have unlocked the secrets of nature, have experimented on material things, and have discovered and created new chemicals and introduced these to human use. In the process, chemistry has changed and helped to mold society. Since chemistry is not only a discipline of the past but is ongoing and vital today, the practicing chemist has a societal impact. The physical and social environment is affected by the work of the practicing chemist. Along with a freedom to unlock the chemical secrets of nature goes a responsibility to use the acquired knowledge properly.

Even though the practicing chemist can, up to a point, separate his or her personal life and professional life, certain societal responsibilities fall within each sphere. The chemist as a professional bears societal responsibilities to colleagues, to management of the institution where he or she works, and to the technical staff; to colleagues through honesty in data collection and presentation and fair criticism of others' work; to management through frankness in communication as to possible harm from current research practices; and to technical staff through a spirit of teamwork and proper credit for their part in research efforts. These professional responsibilities have both interpersonal and societal character.

Even the use of chemical equipment and laboratories has an impact on society. Labs cost money to operate; they need repairs; equipment needs maintenance and spare parts. Chemicals must be properly shipped, stored, and disposed of. Lab emissions may foul the air, water, and land to a noticeable degree. The total research system—personnel, lab, equipment, and material—has economic and societal impacts. The allocation of funding, quality of laboratory safeguards, and emphasis on one research topic or another impact on both the scientific and non-scientific community and include societal factors.

The practicing chemist also bears a societal responsibility with regards to the fruits of his or her labor. It is not enough to discover or to synthesize a new chemical compound; the chemist must be concerned about the proper use or potential abuse of the chemicals produced.

Public spirited chemists are well aware of the following reports:

- The National Cancer Institute and the World Health Organization believe that a high percentage of cancer is environmentally induced. Among the general population the chances of dying from cancer are one in five, and the chances of developing some form of it are one in four.
- Approximately one-third of workers' health problems are caused by exposure to toxic substances in their environment, and at least 200,000 to 500,000 illnesses and 100,000 deaths are caused by occupational disease.
- More than 200,000 infants are born with physical or mental damage each year. About 20% of all birth defects are believed strictly the result of environmental factors such as drugs, chemicals, or radiation, and another 60% are believed to be the result of an interaction of environmental and hereditary factors.
- Fluorocarbons and aerosol spray products continue to concern scientists because of their apparent ability to destroy the ozone layer in the upper atmosphere, thus leading to increased incidence of skin cancer.
- Mercury, lead, and cadmium can attack the central nervous system; carbon tetrachloride and chlorinated phenols can destroy the liver; ethylene glycol and cadmium sulfate produce kidney disease; asbestos and beryllium lead to lung disorders; and lead poisoning can cause mental retardation.

No chemist can be unaffected by these reports. Some chemists wish to discount the impact of the extensive list of chemical hazards, so they concentrate on one or two misnomers or inaccuracies by health and environmental advocates. Other chemists might assert that to some degree all have contributed to the production, promotion, and consumption of these chemicals. Every chemist, however, bears some responsibility to society for the potential threat to the environment and human health stemming from the 30,000 existing, untested commercial chemicals and the hundreds of new ones introduced each year.

Whatever way the chemist is affected, he or she should consider the following questions:

(1) What is the possible use of this chemical?

(2) Are there secondary uses which might prove harmful to average citizens?

(3) How much of the particular chemical is going to be produced?

(4) Does the process being used in manufacturing the chemical require excessive amounts of energy and scarce resources?

(5) Will the chemical explode or be dangerous in shipment and storage?

(6) Will the chemical's use allow for consumer abuse?

(7) What is the animal and human toxicity of the chemical?

(8) Will laborers be harmed in the manufacturing process?

(9) Are there controlled chemical emissions from the manufacturing plant?

(10) Is the chemical biodegradable?

(11) Is the company making false or misleading advertising claims?

(12) Is the chemical product properly labelled when shipped or sold?

The ordinary practicing chemist may find it impossible to answer many or any of these questions, especially if he or she works in a purely research section of a company or academic institution. He or she is not in promotion or advertising, or chemical engineering or sales, or environmental sciences. In fact, any practicing chemist or other professional person might wish to excuse himself or herself from answering these questions—but is such an excuse justified?

An oil company scientist once said that he was happy his company was reprimanded for some of their advertisements; he said the chemists in his lab were somewhat piqued that management never showed ads to scientists to check for accuracy; they left it entirely in the hands of Madison Avenue. He was convinced that commenting on ads was part of the chemist's work and social responsibility.

The above series of questions is not specifically chemical in nature but is intimately connected to the commercial value and societal impact of the chemical. In fact, these questions are often more important from a societal viewpoint than the color retention, absorbency, or durability of a particular chemical product. Merely knowing the toxicological effects of a certain chemical is not enough. The public has a right to know amounts, uses, and other characteristics of the material. The practicing chemist quite often is the most knowledgeable person to hoist a warning flag about a potentially dangerous compound.

Traditional chemistry programs have offered little in the way of toxicology training. What the practicing chemist usually learns in this field is strictly extracurricular. Still, data derived from toxicological testing, reporting, and disclosure is of utmost importance in judging economic and health effects. Are chemical plant workers aware of the dangers of the materials with which they work? Are consumers aware of potential harm a chemical product presents? Through years of laboratory work chemists learn to handle chemicals with caution; they know that their own lifespans can be shortened by failure to take proper safeguards; they hear that many famous chemists live remarkably short lives, and that the average chemist lives 10 years less than the average medical doctor. But while not willing to work in mercury-filled laboratories or to taste each new organic chemical—as was done in the last century—chemists may still be unwilling to share their own empirically based caution with the general populace.

One cannot predict where and when societal responsibility ought to be applied. A mere use of common sense or learned scientific knowledge is not enough. The chemist must be ready to make value judgments which include the importance of society to his or her professional life. Society is more important than quantity and quality of publications or one's career advancement. Affirming this belief by action is another matter. Often a good way to raise oneself to a higher level of public interest consciousness is to communicate with average citizens about issues of mutual concern.

The average American is immersed in a consumer culture which has encouraged intake of increasing quantities of chemicals. Americans are hooked on chemicals: alcohol, hard drugs, over-the-counter drugs, prescription drugs, aerosol sprays, household chemicals, pesticides, and a host of other products. Instead of analyzing the underlying factors that cause overdependence on chemicals, Americans are apt to focus on one chemical which causes harm or potential harm—a food additive or a detergent—but to neglect the whole picture. Chemists can offer valuable assistance in educating the public to respect all chemicals, not just those publicized at one or other time as dangerous.

Water can both give life and cause death by drowning. Salt is necessary for life but is unhealthy in excess. People have to chart the difficult course between what is overly abstemious and what is excessive. This takes thought and education. The small band of public interest advocates are too overworked to perform this task alone. Practicing chemists can help with this education by:

(1) Alerting consumers to overuse of certain items, even before toxicity has been firmly established.

(2) Discussing with citizens substitutes for potentially dangerous consumer products.

(3) Working with PTAs and civic and church groups on drug problems.

(4) Working with those who have no sympathy with drug abusers so they can be led to see that their own overuse of chemicals is a source of the problem.

(5) Supporting local consumer and environmental organizations, offering them advice and encouragement.

(6) Assisting as technical aids in making film strips, educational games, and informational packets on hazardous wastes, indoor pollutants, and natural and synthetic toxins.

(7) Joining a toxicology study group.

(8) Talking with reporters who want to cover a news story about a toxic material. So often chemists want nothing to do with those who might misquote them; however, the risk of misrepresentation is much smaller than the ever-present risk of misinforming the public.

(9) Teaching a course in public interest chemistry at the local high school or community college.

However valuable the educational work on an individual level is in raising the chemists' social consciousness, it is not nearly as important as extending one's societal responsibility to the level of public interest action. The chemist could:

(1) Petition the Federal Trade Commission for proper labelling of chemicals.

(2) Support toxic substance legislation on state and national levels.

(3) Petition for complete corporate disclosure of medical health records to the individuals concerned.

(4) Serve on advisory boards for one's state or federal representative.

Societal responsibility can be fostered through consumer education and political and public interest action. Experience in exercise of this responsibility prepares one to view the benefits and the risks of certain chemical practices in a prudent manner. I say prudent instead of unbiased because all human beings, even scientists, have biases which must be recognized as part of one's subjectivity. No one makes valueless judgments, nor is there such a thing as a purely objective risk-benefit analysis.

The balancing of risks and benefits and the placing on the proper party the burden of proof about toxic substances can create a number of complex legal, economic, and social problems. In a journal editorial Albert F. Plant said (*1*):

> I think the heaviest burden of proof should lie with those who will profit from a new development, not with those who will be exposed to it, but the decision should still try to balance the benefits versus the risks.

Creating a balanced social judgment means weighing risks and benefits. What one person views as a benefit, however, may be a true benefit for only a small portion of the people. Energy "needs" are often extensions of a waste ethic which has profit as the motivating force. Chemical "needs" are often convenience items which are potentially harmful when applied. A remote benefit must be weighed against equally remote risks. The trouble is that the practicing chemist is often immersed in the judgments which have already gone into determining a so-called commercial benefit. While safety cannot be absolutely proved, neither can absolute proof of harm be determined prior to use. Society must make value judgments, and that is what technology assessment is all about.

Recently considerable attention has been paid to the limits to growth in both population and ecomonic areas. Little has been said, however, about limiting chemical research or production. Are there, perhaps, times when chemicals simply should not be produced because society is unprepared to use them properly? The beginnings of such limitation on research may be emerging already in the biological sciences. Research in recombinant DNA has led to concern that researchers might inadvertently allow bacteria bearing new and unusual genetic combinations into labs where they might produce adverse effects on plant, animal, and human populations. A series of committees has been created to produce research guidelines and to protect against mishap.

Perhaps a few specific recommendations are in order. The first is that the ACS establish a committee of responsible scientists to look into the question of limits for chemical use and research. One such candidate for a proscribed list is β-naphthylamine. It would be within the mandate of this committee to treat questionable compounds and production methods and to recommend to the manufacturer less harmful substitutes. For the health and eventual economic viability of the entire chemical profession, a policing of chemical manufacturing methods and products is imperative.

A second recommendation is that the ACS spell out that part of one's professionalism is mandatory disclosure of information about toxic and harmful effects of new or currently used chemicals. It is not enough for a scientist simply to know certain information which is of societal concern. It is imperative that the information go beyond company files and scientific journals and enter the domain of public interest. The art of delivering this information may at times have to include whistle-blowing.

As a member of a professional group, the chemist is becoming conscious of the need to champion good hiring practices, working conditions, job security, and pension policies. One finds this awareness broadening to include remote topics such as the politics of securing grants. When one realizes, however, that it is the public's money at stake, efficient use of resources demands accountability from all, including the academic researcher. Thus, as a third recommendation, the ACS should develop guidelines for treating and exposing the ethical practices dealing with grantsmanship and help to guarantee that society's financial resources are used for the most intellectually and humanly beneficial research proposals.

I hope that these problems, recommendations, and strategies for action incite public spirited chemists to become more concerned about their own societal responsibilities.

Literature Cited

1. Plant, A. F., *Chem. Eng. News* (March 15, 1976).
2. Fritsch, Albert J., "The Contrasumers: A Citizen's Guide to Resource Conservation," Praeger, New York, 1974.

Bibliography

American Academy of Arts and Sciences, Conference on "Scientists in the Public Interest: The Role of Professional Societies," Alta, Utah, Sept. 7–9, 1973.

American Association for the Advancement of Science, Symposium on "Ethics and the Corporate Scientist," Boston, Feb. 19, 1976.

American Physical Society, Symposium on "Secrecy," Washington, D.C., Apr. 24, 1973.

Baram, M. S., "Technology Assessment and Social Control," *Science* (1973) **180**, 465.

Barbour, I. G., "Science and Secularity: The Ethics of Technology," Harper & Row, New York, 1970.

Brooks, H., "Technology and Values: New Ethical Issues Raised by Technological Progress," *Zygon* (1973) **8**, 17.

Brown, M., Ed., "The Social Responsibility of the Scientist," The Free Press, New York, 1971.

Callahan, D., "The Tyranny of Survival," Macmillan, New York, 1973.

Coates, J. F., "Why Public Participation Is Essential in Technology Assessment," *Pub. Adm. Rev.*, Jan./Feb. 1975, pp. 67–69.

Cohen, M. L. and Stepan, J., "Literature of the Law-Science Confrontation, 1965–1975," *Program on Public Conceptions of Science Newsletter,* June 1975, #12, pp. 28–54, Jan. 1976, #14, pp. 32–84.

Cranberg, L., "Science, Ethics and Law," *Zygon* (1967) **2**, 262.

Davis, B. D., "Science, Objectivity, and Moral Values," *Program on Public Conceptions of Science Newsletter,* Apr. 1975, #11, pp. 39–47.

Dror, Y., "Scientific Aid to Value Judgment," *Proc. of the Second International Conference on the Unity of the Sciences,* Tokyo, Nov. 18–21, 1973.

Dubos, R., "Reason Awake: Science for Man," Columbia University Press, 1970. "A God Within," Charles Scribner's Sons, New York, 1972.

Edsall, J. T., "Scientific Freedom and Responsibility," *Science* (1975) **188**, 687.

Feinberg, J., "Doing and Deserving: Essays in the Theory of Responsibility," Princeton University Press, 1974.

Gingerich, O., "The Nature of Scientific Discovery," Smithsonian Institution Press, Washington, 1974. Contains addresses from the forum on "Science and Ethics" in Washington, Apr. 1973.

Henderson, H., "Philosophical Conflict: Reexamining the Goals of Knowledge," *Pub. Adm. Rev.*, Jan./Feb. 1975, pp. 77–80.

Jones, H. W., Ed., "Law and the Social Role of Science," Rockefeller University Press, New York, 1966.

Juergensmeyer, M., "The Ethics of Secrecy," *Ethics and Policy.* Center for Ethics and Social Policy, Apr. 1976.

Kornhauser, W., "Scientists in Industry," University of California, Berkeley, 1962.
Lepkowski, W., "The Limits to Cancer Control," *Nature*, in press.
Lonergan, B. J., "Insight: A Study of Human Understanding," Philosophical Library, New York, 1957.
Lowrance, W. W., "Of Acceptable Risk: Science and the Determination of Safety," William Kaufmann, Inc., Los Altos, 1976.
Meyers, C. J. and Tarlock, A. D., "Selected Legal and Economic Aspects of Environmental Protection," Foundation Press, Mineola, 1974.
"Mount Carmel Declaration," Technion-Israel Instiute of Technology, Dec. 21, 1974.
Nader, R. et al., Eds., "Whistle Blowing: The Report of the Conference on Professional Responsibility," Bantam, New York, 1971.
National Academy of Sciences, "Experiments and Research with Humans: Values in Conflict," 1975. "Principles for Evaluating Chemicals in the Environment," 1975. "Decision Making for Regulating Chemicals in the Environment, 1975.
Perlman, D., "Science and the Mass Media," *Daedalus*, 1974, Summer, 207.
Polyanyi, M., "Meaning," Chicago Press, 1975. "Personal Knowledge: Towards a Post-Critical Philosophy," University of Chicago Press, 1974.
Price, D., "Money and Influence: The Links of Science to Public Policy," *Daedalus*, 1974, Summer.
Primack, J. and Von Hippel, F., "Advice and Dissent," "Scientists in the Political Arena," Basic Books, Inc., New York, 1974.
Pugwash Symposium on Science and Ethics, "International Code of Ethics for Scientists," Jan. 8–10, 1976.
Rabinowich, E., "Back into the Bottle?" *Science and Public Affairs*, Apr. 1973, pp. 19–23.
Ravety, J. R., "Scientific Knowledge and Its Social Problems," Oxford University Press, London, 1971.
Rokeach, M., "Convergence and Divergence between the Value Images of Science and the Values of Science," AAAS Workshop on the Interrelationships between Science and Technology, Apr. 10–12, 1975.
Rosenfeld, A., "Who in Government Is Watching?" *Saturday Review/World*, Nov. 30, 1974, 49.
Rubenstein, E., "Technology and the Public," *IEEE Spectrum 12*, Jan. 1975, pp. 58–64.
Shils, E., "The Confidentiality and Anonymity of Assessment," *Minerva xiii* (1975) **2**, 135.
Sills, D. L., and Gates, R. A., "Environmental Decision Making," *Social Science Research Council Annual Report 1973–1974*, New York, pp. 13–20.
Swartz, E. M., "Hazardous Products Litigation," Lawyers Cooperative Publishing/Bancroft-Whitney, Rochester/San Francisco, 1974, p. 416.
Thring, M., "A Hypocratic Oath for Applied Scientists," *New Sci.* (1971) **49**, 25.
Ziman, J., "Public Knowledge: The Social Dimension of Science," University Press, Cambridge, 1968.

RECEIVED August 19, 1976.

Discussion

Q. What do you think of the new NSF Science For Citizens Program?

A. I am not sure what the impact of the new Science for Citizens Program will have. I think this science–citizen development within the government and within professional groups is going to create a new climate on the part of both ordinary citizens and scientists, especially with respect to ethical and moral scientific problems. Both ordinary citizens and scientists will begin to see how scientific research impacts on our society.

The program has published about two inches of documentation. While I haven't read it all thoroughly, it does seem to contain good material. What will happen now that the documentation has been gathered is anyone's guess. I think the climate for public interest science is changing rapidly both in Washington and throughout the United States. Scientists and citizens are beginning to ask some very serious questions. That is especially exemplified by recent input in the form of testimony on certain bills before Congress which have chemical impact. The time is ripe for a good program.

Q. I wonder if you would comment on what you might view as possible protection for the individual scientist who, when working for an organization, discovers something that he feels is not really in the best interests of society and hasn't yet blown the whistle.

A. Once the situation is understood, the individual scientist should search out a friend, a confidante, and tell him or her the problem and find out what they would do about it. So often one's motives may not be pure. A whistle blower may be a person who might want to leave a company with a bang, and this is the opportunity. The person may have a grudge against someone in the organization. In order to discern one's motivation one needs to talk the matter over with a disinterested party.

The next step deals with whether the problem can be resolved within the organization without whistle blowing or could it be resolved within one's scientific professional organization. If either will work, then that is the proper place. Again discernment is necessary. The whistle blower might get fired or get his or her funding removed. The risk is great.

The next step (granting good motivation and inability to act within traditional structures) is to choose a course of action which has maximum impact. Pick a responsible public media expert to carry the message. Make sure the news gets to the proper people. It is extremely important to choose the proper place and the proper time. Perhaps the choice must be a period when there is a low input of other news. Get it to where the people will read it. If possible do it in association with the professional society so that there will be proper support in case of a threat. Many of the points on whistleblowing have been spelled out in currently available books (2).

10

Affirmative Action and Equal Employment Opportunity

GERALD A. BODNER

Albert Einstein College of Medicine, Yeshiva University, Bronx, N.Y. 10461

The meaning of affirmative action to provide equal opportunity employment is clarified with respect to what the law requires. Although the definition of affirmative action is relatively straightforward—to undertake reasonable efforts to provide equal employment opportunity—problems in interpreting government regulations and defining terms such as "goals," "minorities," and "merit" complicate matters. The author suggests non-governmental means of regulation in dealing with particular aspects of these problems.

In dealing with affirmative action or any other obligation arising out of law and administrative regulations, it is important to distinguish between our legal obligation on the one hand and any additional undertakings an individual or entity might wish to assume beyond the requirements of the law on the other. Here I set forth the former and provide a brief outline for the lay reader of our primary legal obligations arising out of what is referred to as "affirmative action" (*1*).

It is, in my opinion, essential for intelligent decision making that those charged with that responsibility be accurately told what it is they are obliged to do—as opposed to what it is they are being asked to do beyond the pure requirements of the law. There are those who would use what they seem to regard as assumedly desirable legal theories as a bludgeon to force social policies that may or not may be desirable on their own merits. It is, however, neither my intention nor my role to comment on those social policies. I simply wish to make clear what an intensive study of affirmative action laws, regulations, and cases convinces me is the nature of our legal obligation. I leave to others what they may feel equally strongly are our obligations beyond the law.

Let me begin by telling you what, in my legal opinion, affirmative action does not and cannot lawfully require. It does not require an institution to hire a certain percentage or number of individuals from any given ethnic group or sex, even if that institution currently utilizes a lower percentage of such individuals than the percentage supposedly available in that job grouping (2). It does not mean that you are required—or even permitted—to modify standards to increase utilization of members of minority groups or women so long as your existing standards for selection are reasonably job-related and applied equally. It does not mean that you have to allocate to any group of individuals proportional percentages of future promotions, tenured positions, or salary levels. Further, inherent in all this, it does not mean that the government must be made a working partner in determining how you run your institution and meet its legitimate operational and academic needs so long as these decisions include reasonable affirmative action efforts to ensure nondiscrimination.

General Requirements

What affirmative action does mean can probably be best understood if you think of it not so much as an end in itself but as a means towards a very important end. Affirmative action is an obligation to undertake reasonable, good-faith efforts toward achieving the goal of equal employment opportunity. Thus, what is currently the most essential aspect of affirmative action requires that you expand your recruitment sources so that reasonably qualified individuals from a broad array of ethnic groups and both sexes have an opportunity to know of an available position and be considered on the basis of valid and equally applied criteria. Essentially, that means that if your only current method of recruiting a chemist is to call the chairman of the chemistry department at Harvard and say, "who do you know that you can recommend" (something the government labels as the "old boy network"), affirmative action obligations would require you now to expand your recruitment sources on the assumption that the chairman at Harvard is not as likely to know potentially qualified minority or female applicants. Perhaps that means the position must be advertised in some publication. If so, the publication chosen should have a reasonable likelihood of making the position known to potentially qualified minority and female applicants. It does not mean that you cannot continue to contact your colleague at Harvard and ask him for his recommendation, and it does not mean that you cannot fill the job until various minority and female applicants apply. What it does mean is that in addition to your old method of recruiting, you must utilize others so that there is a reasonable opportunity for all groups to know about the position and be considered fairly for it. As long as the means you adopt are reasonably designed to that end, that, in my opinion, is what is required, even if it doesn't result in a single minority or female applicant, or, if there is a broad array of candidates, even if the candidate hired is neither a minority member nor a female.

Goals

Is it necessary to state goals? My legal opinion would be yes. (I base my opinion on the regulations of the U.S. Department of Labor and various enforcement agencies such as H.E.W. as well as the few judicial decisions that have thus far been issued.) However, one must understand what goals are or, at least, must be under the law. Conceptually, in the absence of an affirmative action requirement to set goals, the burden of proving discrimination would lie with the government. Even if not a single black is hired in 10 years, a black applicant would have no relief unless he could prove his rejection was demonstrably the result of discrimination. By being required to state goals (for example, "within the next three years we will hire five additional blacks as chemists") and by failing to meet those goals, what happens in legal terms is tantamount to shifting the burden of proof from the government to you so that you must show that your failure to reach the stated goal is not the result of discrimination. To put it another way, it creates a *prima facie* case against the institution but not an irrefutable one. Thus, the burden of showing non-discrimination can be met by showing that you engaged in recruitment efforts reasonably designed to attract qualified minority candidates who perhaps nevertheless didn't apply, or that despite application of minority and non-minority candidates, you selected a non-minority candidate because he or she was demonstrably the best qualified for the particular job or jobs in question.

Precisely what does goal setting involve (although preciseness has regrettably not been one of the outstanding characteristics of this process)? Conceptually, one sets a goal by determining one's current utilization of minorities and of women in a particular job grouping and then compares it with the supposed percentage of qualified minorities and of women available for placement in that job grouping within your recruitment area. Then, simply by calculating the rate of expected turnover, one calculates the timetable for reaching the percentage of available minorities and of women, respectively.

Now for the realities. First, determining the available percentage of minorities or women is anything but simple or clear. Is your proper availability pool all individuals with a Ph.D. in chemistry from any American educational institution, or is that unrealistically inadequate for your particular institution since you draw primarily from graduates of Harvard, Berkeley, and Stanford? [The "Guidelines" of the U.S. Department of Labor for educational institutions specifically permit a pool based on the "feeder school" concept, at least for minorities (3).] If your institution is seeking chemistry professors, should the pool be limited to reflect only those Ph.D.'s that have ability or interest in teaching at the required level? Should the pool include only those with employment experience as chemists or also reflect currently unemployed chemists as well? If you're looking for a senior professor of chemistry, should your pool be limited to chemists who received their Ph.D.'s at least five years previously, or must it also include more recent graduates? Then, of course, there are the problems involved in how you define your job grouping, your recruitment area, and your current utilization of rather unspecifically defined minority groups [for example, there is continuing debate over whether persons from India or Pakistan are minority members (*4*)]. The significance of these problems can be vast. Defining the job grouping, for example, involves the basic principle or possible fault inherent in goal setting in that you get no credit for job groupings in which you have overutilization, yet you are, in a sense, penalized for job groupings in which minorities or women are statistically underutilized. Therefore, the question of whether the proper job grouping is, for example, the chemistry department alone, all science departments together, or the entire school encompassing the chemistry department takes on substantial importance. (We recently obtained a determination that the appropriate goal-setting units for the faculty at a college of medicine consisted of a single unit for all clinical science faculty and a single unit for all basic science faculty as opposed to much more finite breakdowns originally demanded by H.E.W.) However, the possible advantages of larger job groupings also involve the inherent inaccuracy of combining availability percentages for varied disciplines and applying that single percentage to disciplines where it may be unrealistic. In short, there are no easy answers and very few precisely defined questions.

Merit

Let me again emphasize that affirmative action legally could not and does not limit the right of any institution to make any and all decisions on the basis of merit—whatever its effect on the percentage of people hired from various groups; whatever different salary levels it results in for people in the same job category, albeit at different levels of ability; and whatever its effect in terms of promotion, tenure, or the like. This, of course, is not to say that merit must be used as a determinant in regard to every decision regarding every job. What is required, however, is that the determinant not be based on ethnic background, sex, or other prohibited criteria. Thus, for example, years of service is a permissible determinant of salary level where this is regularly and fairly used at an institution.

There are two important requirements, however, in regard to job criteria, including those based on merit. First, the criteria used must be validly job related, and secondly they must be equally applied. Thus, you cannot impose a requirement that chemists have a minor in literature where such requirement would have a disparate effect on minority or female applicants unless you can show that requirement is reasonably and properly related to job performance (5). Additionally, you obviously cannot apply a requirement to certain applicants which you do not apply to others, although I would be inclined to believe that you could impose current requirements that were not applied to every present holdover in the job category on the ground that valid job-related standards have been raised.

That aside, the difficult question remains in regard to merit: how do you provide evidence that you have, in fact, based your decisions on merit, particularly when this is offered as a justification for your numerical failure to meet your stated goals. In short, how do you define and explain merit in a rational manner that enables some reviewing mechanism to be convinced that merit is not being used as an excuse for lack of good-faith affirmative action efforts or for discrimination. It is a problem perhaps compounded by a bureaucratic suspicion of any system that cannot be quantified and programmed. However, that obviously cannot be allowed to deter a merit system. Merit is not, as I have sought for long hours to convince government officials, the number of publications a faculty member writes—not even if we factor in (as some government officials have suggested) a weighting for the particular journal in which it is published or the length of the article. Rather, at some point, an evaluation of merit necessarily reflects qualitative and subjective human judgments of worth and ability. How, the government agent asks, do I know that supposed merit is not being used as a cover-up for intentional or unintentional discrimination? (Note that the intention to discriminate—as opposed to the actual effect of discrimination—is not a necessary ingredient in a finding of fault.) It's a valid question for which there are as yet, in the brave new world of affirmative action requirements, only tentative answers. One such possible answer suggested by some government official would involve listing the applicable subjective criteria (excellence in scholarly research, citizenship functions and responsibilities, teaching ability, etc.) and assign these a weighted factor in the decision-making process. Then whoever is responsible—the chairman, faculty committee, etc.—would numerically evaluate each individual's rating in each factor (so that Dr. X gets 8 points in scholarly research, Dr. Y gets 3 points in scholarly research, etc.) and multiply this rating in each criteria by the weight given to such criteria to arrive at a weighted result that can be compared from one individual to another. To be frank, I am not entirely clear how this system is less subjective. While I recognize it may be of some comfort to the government to at least have an employer specify the criteria and assign a weighting to each criteria, the government must still deal with the fact that at some time human judgment must be involved in any merit-based decision.

A second possible way of dealing with these issues, and one that I strongly recommend be explored by many institutions, is really an attempt to deal with the legitimate concern of many institutions of having to justify professional judgments to individuals who lack the expertise in the given scientific field. A possible solution to this is to establish some reasonable non-governmental means for reviewing decisions. Thus, if the chairman decides that the salary of Dr. Mary Jones is to be $10,000 and that of Dr. John Jones is to be $20,000, and Mary claims that decision is based on sex discrimination, there should be some forum (perhaps three noted professors in chemistry from within and without the institution) who can intelligently review the professional basis for the chairman's judgment. This probably would involve hearing from both the chairman and Mary, among others, and considering the professional, judgmental considerations involved in the decision. I strongly suggest that the establishment of a non-governmental review system by experts in the given field be seriously considered. While it is clearly not an absolute bar to governmental review, there is reasonable likelihood of its decision's being given significant weight in either an administrative or legal review.

Summary

Affirmative action is still a developing area of regulation with many of its precise requirements not yet definitively resolved in either an administrative or judicial sense. I strongly suggest that in dealing with specific problems and questions on affirmative action at your institution, you first obtain the opinion of your counsel concerning the precise requirements of affirmative action and then make the necessary decision with an understanding of what affirmative action actually does and does not require, as well as what is otherwise best for your institution.

Literature Cited

1. *Executive Order 11246*. Further defined by *U.S. Department of Labor Regulations* 41 C.F.R. 60-1.1 et seq.
2. Ibid., 60-2.12(e).
3. "Memorandum to College and University Presidents," Aug. 1975, U.S. Department of Labor.
4. *U.S. Department of Labor Regulations* 41 C.F.R. 60-1.3.
5. Griggs vs. Duke Power Co., 401 U.S. 424 (1971).

RECEIVED November, 1976.

11

Tax Effects of Retirement Plans

FRANK PAUSCH

Internal Revenue Service, P.O. Box 709, Church St. Station,
New York, N.Y. 10008

Setting up one's own retirement plan holds two advantages: a deduction may be made against the current income, and the money appropriated under the plan remains tax-exempt for a specified period of time. Self-employed individuals may choose from two plans, an individual plan or the Keogh (HR–10) plan. Those not self-employed may only establish an individual-type retirement plan. These plans are governed by specific requirements for eligibility and certain restrictions applied when the plan is in effect.

In considering the various tax effects of certain retirement plans which are available to members of the chemical profession, there are two major advantages in establishing a retirement plan. The first advantage is a current deduction which can be made against taxable income in the year in which the plan is established. The second advantage provides that any earnings which are generated by contributions made under these plans over the years remain tax exempt until distributions are made, generally at retirement. Two types of plans are available to chemists, the appropriate plan depending upon an individual's situation. A chemist is considered either an employee or a self-employed individual. Those who are classified as employees are entitled to establish an individual retirement account. For those who are classified as self-employed individuals, that same individual retirement account might be available, but those individuals may, in lieu of establishing an individual retirement account, be entitled to the tax advantages of a Keogh Plan, also known as an HR–10 Plan.

There are general requirements in establishing an individual retirement account. First you must be an employee or a self-employed individual who is not participating in a qualified pension, profit sharing, or stock bonus plan of an employer. Therefore, if you are employed by a corporation (or a non-corporate entity) and that organization maintains a retirement plan under which you are covered, you are excluded from establishing an individual retirement account. If you are self-employed and you have established a Keogh plan, you cannot set up an individual retirement account. If you are employed by a governmental unit, and that unit provides a governmental plan, you are also precluded from setting up an individual account.

Once it has been established that you may set up an individual retirement account, you must consider the maximum amount of contributions which can be made under this type of plan. The contribution limitations are $1500 or 15% of your earned income, whichever is less. For these type of plans, earned income is defined as wages, salaries, professional fees, and self-employment income. It does not, however, include earnings from property, such as interest, dividends, or rents. These latter types of earnings are considered passive income and cannot be considered in calculating the amount of contribution which may be made.

These contributions, in order to be deductible, must be paid by the last day of your taxable year, December 31 in most cases. For taxable years beginning after 1976, this requirement has been changed so that contributions may be made not later than the 45th day after the end of the taxable year and still be deductible for that taxable year. In addition these contributions must be in cash; they may not be in property. There are times when an individual, by the last day authorized for making such contributions, does not have a complete picture of his total earnings for that year. Therefore, it is possible that he may contribute in excess of 15% of his earned income. If that were to occur, the individual would be able, prior to the due date of his return, to get back such excess contributions, plus interest on such excess, and thereby avoid a 6% excise tax penalty on the excess that was made. However, if the contributions are not returned, the employee is subject to the 6% excise tax penalty.

Once an individual has established these plans, distribution provisions under these plans are restricted to the time at which they may commence. Distributions from an individual retirement account may not begin before an individual reaches age 59½. If distributions are made prior to that age, they are subject to a 10% premature distributions penalty in addition to being included in the individual's taxable income for that year. Furthermore, distributions may not be postponed beyond the taxable year in which the individual attains age 70½. So, in other words, between age 59½ and 70½ distributions may commence. Once distributions do occur, any amounts received under the plan are taxed as ordinary income includable in the individual's gross income for that year and are taxed at ordinary rates—no capital gains or special averaging is permitted.

To establish a Keogh Plan, the individual must be a self-employed individual who has earned income from a trade or business in which he renders personal services. Unlike individual retirement accounts, a self-employed individual may participate in another qualified pension, profit sharing, or stock bonus plan (but not in an individual retirement account) and still establish a Keogh Plan. There are other advantages for a self-employed person in establishing this plan.

The maximum contributions under a Keough Plan are substantially higher than under an individual retirement account. The contribution limitations are the lesser of 15% of earned income or $7500. Earned income for contribution purposes does not include wages, salaries, dividends, or interest. Earned income includes only net earnings from self-employment.

In order for the contributions to be deductible, payments for both cash and accrual basis taxpayer may be made no later than the due date of the income tax return. If a contribution is made in excess of the 15% or $7500 limitations, the excess is subject to a yearly 6% excise tax penalty until such excess contribution is fully utilized. Distributions under a Keogh Plan may not commence before an individual reaches age 59½, except for reasons of disability or death. Like the individual retirement account, distributions must commence no later than the taxable year in which the individual attains age 70½.

When distributions do commence, they may be available for capital gains and/or a special 10-year averaging. Distributions attributable to years of participation under the plan prior to Jan. 1, 1977 are subject to capital gains—for years after Dec. 31, 1973, to ordinary income rates. If certain conditions are met, however, an individual may avail himself of the special 10-year averaging method for the ordinary income portion of the distribution by filing form 4972 with his form 1040. These conditions are specified on form 4972 and should be reviewed if a distribution is received under a Keogh Plan.

RECEIVED March 17, 1977.

12

Chemists and the Federal Tax Law

RICHARD HAAS

Internal Revenue Service, P.O. Box 3100, Church St. Station,
New York, N.Y. 10008

The determination of tax status and the procedures to guide one in determining the IRS thinking on a particular tax status, new rules for office-at-home deductions, other allowable deductions, and patents, nonpatentable secret processes, and copyrights are addressed briefly and generally.

The first subject to be covered is the determination of your tax status as an employee or an independent contractor. Various tax benefits and obligations which depend on this classification are discussed a little later. I discuss this subject only in a general rather than a specific manner because this is a highly complex subject in which certainty is often difficult to come by. There are many factors to be considered in determining status, and the answer depends not on any one factor but on a balanced picture of all the factors contained in any one situation. The general legal criteria is whether or not the employer has the right to direct and to control the manner and method of performing the service.

At one pole you have the clearly independent free agent who operates his own lab, makes his services available to the public, works for several clients, schedules his work as he sees fit, and hires his own assistants. At the other pole you have the employee of a major corporation who can be discharged at any time. He is supplied with tools, a place to work, and assistants. He is told in what area and on what project to work and can be moved from one project to another.

Between these extremes you can have an almost unlimited number of combinations of circumstances, and as you approach the middle ground it becomes harder and harder to know into which category you fall. The key factor as mentioned before is direction and control over the method and manner of performing the service. This does not mean, even if you are clearly an employee, that the boss looks over your shoulder and tells you every move to make. As chemists you are professionals and in an employment situation are expected to be able to work independently. It is not the actual exercise of control but the employer's *legal right* to exercise such control which is determining.

Once you are classified, the tax result is more or less fixed. Therefore, it is important to be able to structure your situation in advance. The IRS has three administrative procedures whereby you can get a prior indication of the IRS thinking and then structure your situation to ensure the desired result. I would recommend the use of these procedures.

The first method is to request an Information Letter from your local IRS office. Although this letter is not binding on either party, it is valuable as an indication of IRS thinking on the matter. I recommend this procedure because it is potentially the quickest and involves the least red tape. The other two procedures are known as Determination Letters and Rulings. Both are formal proceedings that are binding on both parties and generally take some time to issue.

You should be aware that if you are involved in two or more situations, you could be an employee in one and an independent contractor in another. As an employee, both social security tax and income tax are withheld from your salary. As an independent contractor, you must estimate your income tax and social security tax liability and pay it in advance quarterly installments. Therefore, until the end of the year the amount withheld on salary will generally exceed the amount paid in by the independent contractor, and based solely on the value of the use of money, an independent contractor theoretically saves the interest on this difference. Also, for 1976 the amount of social security withheld from your salary is 5.85% of the first $15,300, but the amount of social security tax an independent contractor must pay is 7.9% of the first $15,300 of self-employment income. This difference in percentages can result in a maximum of $313 savings for the employee over the independent contractor.

Many of you probably maintain an area of your home for use as a second office. The cost of maintaining such an area may be deductible. The problem in this area is that one section of the code says that no deduction shall be allowed for personal, living, or family expenses, while another section of the code provides a deduction for all the ordinary and necessary expenses of carrying on a trade or business. You can see that these two sections can conflict where you use a part of your personal home as a business office. On the one hand, the IRS is concerned that no deduction be allowed for personal expenses, and on the other hand the taxpayer is concerned that he be allowed a deduction for his legitimate business expenses.

Before 1976, home office expenses were deductible if the conditions of employment necessitated work at home, the home office was used on a regular basis, and the regular business office was not available at the same time the employee used the home office. The Tax Reform Act of 1976 substantially restricted the office-in-the-home deduction. For tax years beginning in 1976 no deduction is allowed for an office in the home unless such office is used *exclusively* and on a regular basis as either your *principal* place of business or as a place where you *regularly* meet with clients or customers in the normal course of your business. In addition to the above requirements, if you are an employee, the business use of the home must be for the convenience of your employer. Included in the deduction for office at home is a pro-rata share of depreciation if you own your home, a share of light, heat, and power, and a share of rent if you are a renter.

As either an employee or an independent contractor you are entitled to deduct educational expenses if they are incurred to maintain or to improve your skills in a trade or business in which you are already engaged. You may *not* deduct educational expenses that will qualify you for a new trade or business. Costs incurred in inventing may be deducted if you are in the business of inventing and you make a special election under code section 174 to deduct your costs. If you do not so elect, such costs must be capitalized and can be deducted only ratably over the useful life of the invented item. If you are in the business of writing, costs incurred *must* be capitalized and can be written off only over the useful life of the book. As either an employee or independent contractor you are entitled to deduct trade publications, memberships in professional organizations, and the costs incurred in attending a convention.

One final word about deductions on your tax return. It is imperative that you keep adequate records. Without them you might not be allowed any deduction at all. It is your responsibility to prove that you are entitled to a deduction. Another area in which you as a chemist might become involved with the federal tax law is the area of patents, secret processes which are not patentable, and copyrights. Under general rules, when you sell outright something which you own, you realize capital gain taxable at capital gain rates, but if you merely license its use, you have royalty income taxable at higher than ordinary income rates.

When Congress considered the question of capital gains for patents and copyrights, it altered this general rule, greatly favoring patents over copyrights although the two items are conceptually quite similar. A law was passed making it impossible for an author to get capital gain from the sale of a copyright but extremely easy for an inventor to get capital gain from a patent. In fact, even if you transfer a patent under a contract calling for periodic payments tied to the amount of the patented product that is sold, which on its face looks like a license rather than a sale, you can still get capital gain. All you must do is be sure that you have disposed of all substantial rights in the patent. This means that you cannot retain any valuable rights after the transfer. The question of what is a valuable right has produced much litigation, which we cannot discuss here.

If, in an employment situation, you come up with an invention and voluntarily assign the patent to your employer, you will get capital gain if the employer's payments to you are not compensation. Payments will generally be considered compensation if you have been hired to invent. If you have not been hired to invent but are required, under a standard form employment contract, to assign the patent to the employer, you might get capital gain. The answer depends on whether or not the extra money received over regular compensation is intended as compensation or as payment for the patent. The issue is inherently factual and could very possibly lead to litigation.

The treatment of nonpatentable secret processes is not covered by any one specific statutory code section as in the case of patents, but it is governed by all the rules generally applicable to capital transactions. While you can't confidently predict results in this area, it is possible to get capital gain treatment from the sale of a nonpatentable secret process. Such a result could very possibly be challenged by the IRS on the grounds that the transaction is a license or that the secret process is not a capital asset.

I have only tried to give you a broad overview to make you aware of some of the tax aspects inherent in patents and copyrights. There exists in the area of patents tremendous opportunities for creative tax planning which are not limited merely to the benefits of getting capital gain rather than ordinary income. I would strongly recommend consultation with a tax attorney if you find yourself in a patent situation that could generate significant amounts of income.

RECEIVED April 12, 1977.

INDEX

INDEX

The text of this book is set in 10 point Caledonia with two points of leading. The chapter numerals are set in 30 point Garamond; the chapter titles are set in 18 point Garamond Bold.

The book is printed in offset on White Decision Opaque, 60-pound. The cover is Joanna Book Binding blue linen.

Jacket design is by Linda Mattingly.
Editing and production by Kevin C. Sullivan.

The book was composed by Service Composition Co., Baltimore, Md., printed and bound by The Maple Press Co., York, Pa.